Industrial Noise Control

Paul N. Cheremisinoff

PTR Prentice Hall
Englewood Cliffs, New Jersey 07632

Library of Congress Cataloging-in-Publication Data

Cheremisinoff, Paul N.
Industrial noise control / Paul N. Cheremisinoff.
p. cm.
Includes bibloigraphical references and index.
ISBN (invalid) 0-13-466202-1
1. Noise control. 2. Vibration. 3. Industrial noise. I. Title.
TD892.C437 1993
620.2'3--dc20 92-41864
CIP

Editorial/production supervision: *Brendan M. Stewart*
Buyer: *Mary Elizabeth McCartney*
Acquisitions editor: *Michael Hays*

A Simon & Schuster Company
Englewood Cliffs, New Jersey 07632

The publisher offers discounts on this book when ordered in bulk quantities. For more information, contact:

Corporate Sales Department
PTR Prentice Hall
113 Sylvan Avenue
Englewood Cliffs, New Jersey 07632

Phone: 201-592-2863
Fax: 201-592-2249

Printed in the United States of America
10 9 8 7 6 5 4 3 2 1

ISBN 0-13-466202-4

Prentice-Hall International (UK) Limited, *London*
Prentice-Hall of Australia Pty. Limited, *Sydney*
Prentice-Hall Canada Inc., *Toronto*
Prentice-Hall Hispanoamericana, S.A., *Mexico*
Prentice Hall of India Private Limited, *New Delhi*
Prentice-Hall of Japan, Inc., *Tokyo*
Simon & Schuster Asia Pte. Ltd., *Singapore*
Editora Prentice-Hall do Brasil, Ltda., *Rio de Janeiro*

CONTENTS

PREFACE

1 ***NOISE EFFECTS*** *1*

THE HUMAN EAR 2
THE EFFECTS OF NOISE ON THE HUMAN EAR 4
TOLERANCES 4
DANGEROUS PROPERTIES OF NOISE 4
NOISE AFFECTS THE MIND AND OUTPUT 6
EFFECTS ON THE HUMAN BODY 7
AUDIOGRAMS 8
EFFECTS OF VIBRATION ON MAN 8

2 ***NOISE REGULATION*** *11*

OCCUPATIONAL SAFETY AND HEALTH ACT (OSHA) 11
Responsibilities of Employers 12
Responsibilities of Employees 12
Administration 13
OSHA Standards 13
OSHA'S REGULATIONS ON NOISE CONTROL 18
Impulse or Impact Noise 19
Variable Noise 19
Engineering Control Measures 19
Administrative Controls 20
WALSH-HEALEY PUBLIC CONTRACTS ACT 21
PRIMER ON OSHA 22
Purpose of the Act 22
Businesses Affected 22
Duties of Employers 23
Duties and Rights of Employees 23
Occupational Safety and Health Standards 24
Administration 24

Complaints of Violations 24
Inspections 25
Notification of Proposed Penalty 25
Contesting a Citation 25
Time for Abatement of Hazards 26
Variances 26
Failure to Correct Violations within Allowed Time 26
Penalties for Violations 26
Recordkeeping Requirements 27
Statistics 27
Imminent Dangers 27
Protection Against Harassment 27
What an Employer Should be Doing 27
GUIDE FOR IMPLEMENTATION 28
Policy 28
Procedures 28
Recordkeeping Requirements 29
NOISE ENFORCEMENT DIVISION 30
U.S. ENVIRONMENTAL PROTECTION AGENCY 30

3 *ACOUSTICS* 32

NATURE OF SOUND 32
THE SOUND FIELD 35
SCALES FOR MEASUREMENTS—THE DECIBEL 38
LOUDNESS AND OBJECTIVE MEASUREMENTS 39
COMPLEX NOISE 42

4 *ENGINEERING CONTROLS AND DESIGN* 44

MAINTENANCE 44
MACHINES SUBSTITUTION 46
PROCESS SUBSTITUTION 47
VIBRATION DAMPING 47
EXCESSIVE VIBRATION CAUSES 48
ISOLATION AND DAMPING OF VIBRATION 50
INCLUSION OF NOISE LEVEL SPECIFICATIONS WHEN ORDERING NEW EQUIPMENT 52
ISOLATING NOISE SOURCES 52
OPERATOR ISOLATION 54
SYSTEMS DESIGN 54

5 *PERSONAL SAFETY* 57

ACOUSTIC PROBLEM 57

EAR PROTECTION 59
Insert Type 61
MUFF-TYPE PROTECTORS 62
COMBINED USE OF MUFFS AND INSERTS 64
COMMUNICATION WHILE WEARING EAR PROTECTORS 64
EAR PROTECTION PROGRAM 66

6 *ENCLOSURES AND BARRIERS* *68*

LEAD AS A NOISE BARRIER 68
PLENUM BARRIERS 70
HOW TO INSTALL PLENUM BARRIERS 71
BARRIERS AROUND PIPES AND WIRES 72
BARRIERS AROUND RECTANGULAR DUCTWORK 73
BARRIERS UNDER TWO-WAY JOIST SYSTEMS 78
HIGH TRANSMISSION LOSS CEILINGS 79
ADDITIONAL SOUND-CONTROL MATERIALS 80
ACOUSTICAL FOAMS 80
FOAM WITH POLYMER FILM SURFACE 82
ACOUSTICAL PANEL 82
POLYURETHANE FOAM 84
NYLON IN NOISE REDUCTION 87
Nylon Gears 87
PVC FILM 88
DAMPING COMPOUND 89
SILENCERS AND SUPPRESSOR 90

7 *VIBRATION* *93*

NATURE OF VIBRATION 93
Periodic Functions 93
Random Vibration 95
Transient Vibration 96
SOURCES OF VIBRATION 96
VIBRATION EFFECTS 97
VIBRATION MEASUREMENT 97
DISPLACEMENT, VELOCITY AND ACCELERATION 99
PEAK, AVERAGE, AND RMS VALUES 103
PRINCIPLES OF VIBRATION TRANSDUCERS 104
Displacement Measurement 104
Velocity Measurement 106
VIBRATION METERS 109

8 VIBRATION CONTROL *111*

PADS AND MOUNTS 111
RUBBER ISOLATORS 112
SPRING SYSTEMS 113
Air Springs for Vibration Control 113
ADJUSTABLE DAMPERS 114
TYPES OF VIBRATION MOUNTS 116
BASIC INSTALLATION PRINCIPLES 117
SEVERAL TYPES OF INDUSTRIAL MOUNTINGS 119
SHAFT COUPLINGS 126
VIBRATION PLATES 127
BLOCKS 128
Some Applications for Isolators 128

9 VENTILATING SYSTEMS *131*

Ventilating Systems 131
Air-Conditioning Systems 132
NOISE CRITERIA 133
Loudness and Loudness Level 133
Speech Interference Level 133
NC and PNC Ratings 133
Noise Pollution Level (NPL) 135
A-Weighted Sound Level (L_A) 135
Insertion Loss 135
Transmission Loss (L_{TL}) 135
Dynamic Transmission Loss 135
MEASUREMENT TECHNIQUES 136
NOISE REDUCTION 137

10 INSTRUMENTATION *141*

MICROPHONES 141
SOUND LEVEL METERS 146
CALIBRATION 150
THE ACOUSTICAL MEASUREMENT 151
DOSIMETERS 151
FREQUENCY ANALYZERS 152
AMPLITUDE DISTRIBUTION ANALYZERS 153

11 AUDIOMETRIC TESTING AND DOSIMETERS *156*

AUDIOMETERS 156
ACCURACY 157

RECORDS 158
AUDIOMETRIC TEST BOOTHS 158
ALTERNATE PROGRAMS 160
DOSIMETERS 161

12 INTERPOLATION AND MAPPING *162*

NOISE LEVEL AND FREQUENCY WEIGHTING 162
INDUSTRIAL HEARING DAMAGE RISK 163
COMMUNITY NOISE FROM INDUSTRIAL SOURCES 163
NOISE CONTROL 163
COMPLIANCE WITH CODES AND STANDARDS 163
NOISE-LEVEL MAPPING 164
POINT SOURCES OF NOISE 164
THE INVERSE SQUARE LAW 165
THE NEAR AND REVERBERANT FIELD 165
PLOTTING POINT SOURCE NOISE LEVELS 166
LINE SOURCES OF NOISE 168
THE INVERSE DISTANCE LAW 168
PLOTTING LINE-SOURCE NOISE LEVELS 169
COMBINED NOISE SOURCES 170
NOISE CONTOUR MAPPING FOR LINE AND POINT SOURCES 172
NOISE CONTOURS FOR COMBINED NOISE SOURCES 173

GLOSSARY **174**

INDEX 186

PREFACE

Controlling noise has been a traditional challenge in the industrial and business world. There are many reasons to be concerned about noise because studies have shown that workers exposed to high noise levels have low working efficiency. Noise can furthermore be responsible for a wide range of tension, heart disorders, and circulatory problems.

This book is intended for use by those faced with noise problems in the workplace. It should be of use to consultants, managers, and engineers, as well as all those concerned with noise control. It is written in technical language to facilitate its use and stresses the practical rather than the theoretical. Regulations and the law require that employers provide a noise control program whenever employees work in an environment exposing them to noise hazards. Hopefully, this book will provide the necessary information for a safer and more productive work environment.

The author is indebted and acknowledges the following individuals who provided information contained in this book and an earlier work by the author:

Anthony J. Schneider	Chapter 3—Acoustics Chapter 7—Vibration Chapter 10—Instrumentation
David Marsh	Chapter 10—Instrumentation
Nicholas P. Cheremisinoff	Chapter 9—Ventilation Systems
E.J. Bonano	Chapter 9—Ventilation Systems
Charles E. Wilson	Chapter 12—Interpolation and Mapping

Major contributions by these experts in the accompanying chapters make this book a comprehensive tool for those concerned with noise.

Since most books on noise are oriented architecturally—not industrially—this book will be especially valuable to those who may have found it difficult to find practical solutions to noise control problems in industry. This includes safety, mechanical, civil, environmental, and chemical engineers, plant and maintenance managers, and industrial hygienists and managers.

Paul N. Cheremisinoff

1 NOISE EFFECTS

Every industry, trade, occupation, and process using equipment, methods, apparatus, or materials that generate noise above certain levels involve elements of danger to the health and safety of employees and other persons frequenting such areas. Medical studies and other sources of evidence indicate that noise having certain characteristics affects the auditory system of the human body, causing loss of hearing. Loss of hearing occurs due to the cumulative effect of exposure to sound above a maximum intensity and over a maximum duration in a given period of time.

When employees are subjected to sound exceeding prescribed levels, employers must institute feasible engineering or administrative controls designed to decrease noise levels in working areas. If given controls fail to reduce noise levels below prescribed levels personal protective equipment must be provided and its proper use by the exposed personnel must be enforced by management. To monitor the effectiveness of the engineering and administrative controls, and the proper use and effectiveness of personal protective equipment, a continuing effective hearing conservation program should be administered.

The three noise control measures may be defined as follows: (1) *Engineering controls* are those which reduce the sound intensity either at the source of the noise or in the hearing zone of the worker; (2) *administrative controls* are those which limit the duration of the workers' exposure to noise levels above 90 dBA to the table of Permissible Noise Exposures; and (3) *personal protective equipment* (ear muffs, plugs, and so on) shall be provided and used to reduce sound levels to permissible levels. It is important to note that the use of personal protective equipment is considered to be an interim measure and is not acceptable as a permanent solution to noise problems.

Surveys indicate that more than half of industrial machines generate noise levels of 90 to 100 decibels. This presents a definite health hazard to workers. Noise-induced hearing loss is dangerous because, once incurred, normal hearing can almost never be fully restored.

Four options are open in arresting this problem. The best is to reduce the noise level at its source. However, this is not always feasible. Reduction of noise generated by the machine may be accomplished by:

- proper acoustical design in machinery
- modifying existing machine design
- muffling
- changing the process entirely

A second approach is to reduce the amount of sound transmitted through the plant. This can be accomplished by:

- increasing the distance between the work vicinity and source
- installing acoustical barriers between the work area and sound source
- mounting equipment on vibration stands to reduce noise transmission through the building

This second list may prove to be costly and ineffective on the noise levels in the immediate area.

The third choice is to revise operational procedures. This may involve:

- changing job schedules,
- rotating personnel,
- providing longer or more frequent work breaks.

This approach cuts down on continuous noise levels; however, it may be uneconomical.

A final approach is to provide the individual with personal ear protection. This seems to be the simplest solution, both economically and to the machine operator; however, there are certain drawbacks. The majority of individuals find personal ear protection devices uncomfortable, irritating and, in many cases, cumbersome.

One of the major problems facing industry today is noise pollution. Data show that over 10 percent of the U.S. work force suffers some hearing impairment due to exposure to high noise levels.

THE HUMAN EAR

The human ear diagrammed in Figure 1-1 can be broken down into three basic parts. The *outer ear* contains the external auditory canal which functionally carries sound waves to the tympanic membrane, or eardrum. The *middle ear* begins with the eardrum which is a strong flexible tissue with a cone-type construction. When sound waves travel through the external auditory canal and hit the eardrum, they will cause it to vibrate. Also, part of the middle

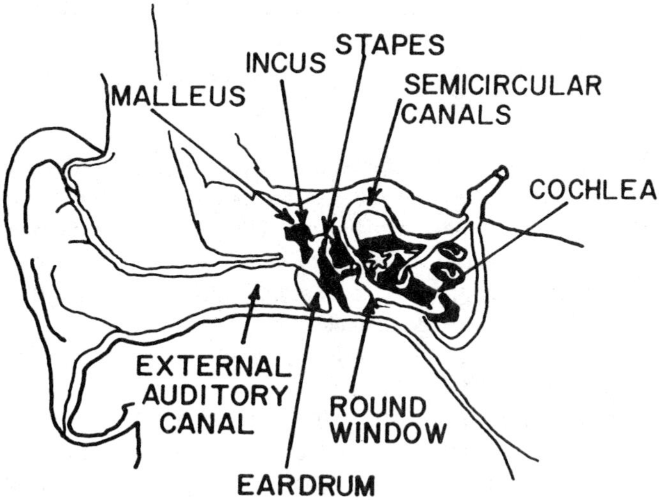

Figure 1-1 Diagram of the human ear.

ear is the malleus or hammer, which is a bone that is connected to the eardrum. The malleus is linked to two other bones which move within a very small space. This assemblage of bones, the smallest in the human body, are known as the ossicles and transmit sound waves to the *inner ear* from the eardrum. They also modify sound by either amplifying or diminishing it to protect the inner car. At the other end of these bones is the foot of thc stapes which directly transmits sound to the *inner ear* and is approximately one thirtieth the area of the eardrum.

We can see that the middle ear regulates the level of sound that enters the external auditory canal, and also protects the inner ear. It is approximately 2 cm^3 in volume, filled with air which dampens the low-frequency rocking of thc ossicles. In addition to this arrangement, there are two muscles attached to the stapes and the malleus which are known as the tensor tympani and the stapedius muscles. The function of these muscles is to tighten the eardrum and the motion of the ossicles, thereby lessening the efficiency of sound transmission. This function is known as the acoustic reflex and is carried out by command of the brain just after every loud sound reaches the eardrum.

The inner ear is a very complex system of boney, fluid-filled crevices lying deep inside the temple. The components which make up the inner ear have nothing to do with hearing but are responsible for our senses. One component is the utricle, which gives us our sense of acceleration and gravity. Another is an arrangement of three semicircular canals which gives us our sense of orientation, space, and balance. The part of the ear with which we actually hear is called the cochlea. When sound waves travel through the external auditory canal, the foot of the stapes bone knocks against the oval window, which is a wide opening in the cochlea, and sound is transmitted to the liquid inside. The round window that lies just below the oval window is an elastic membrane which is the final component that sound waves reach in the human ear.

THE EFFECTS OF NOISE ON THE HUMAN EAR

The ear has its own defense mechanism against noise—the acoustic reflex. However, this reflex has vital weak points in its defenses. First the muscles within the middle ear can become fatigued and slow if overused. In persons who work in an environment with high noise levels, these muscles will gradually lose their strength and thus more noise will reach the inner ear. Second, these muscles can be affected by chemicals within the working environment. Finally, the acoustic reflex is an ear-to-brain-to-ear circuit which takes at least .009 second to perform.

Persons with poor acoustic reflex usually are subjected to temporary hearing loss when they come in contact with a loud noise. Much of the temporary hearing loss caused by noise occurs during the first hour of exposure. Recovery of hearing can be complete several hours after the noise stops. In short, the ear will recover to its full hearing potential after its muscles have had time to rest. However, the period of recovery depends on individual variation and the level of noise which caused the deafness.

TOLERANCES

Persons in different age groups have different tolerances to various noise levels. There have been many investigations made of the threshold of hearing. When young people who have good hearing ability are tested, a characteristic known as the minimum audible field (MAF) is obtained. This characteristic indicates the level of a tone that can just be detected in very quiet surroundings under free-field conditions as a function of frequency of the tone. Under free-field conditions the sound source is equidistant from any objects including the ground. Therefore, the sound pressure is distributed uniformly in all directions, and doubling the distance from the source will cut the sound pressure in half. At low frequencies sound-pressure level must be high before a tone can be detected. At a sound level of about 120 dB a listener would be very uncomfortable. At 140 dB the listener would experience pain.

DANGEROUS PROPERTIES OF NOISE

Noises that pose the greatest threat to the human body are those which are the highest pitched, loudest, poorest in tone, and longest lasting. Another dangerous type of sound—capable of rupturing the eardrum—is the sound of an explosion. However, when only the eardrum is ruptured the inner and middle ear are protected. In time, the eardrum heals and full hearing is usually restored.

Deafness due to noise usually occurs in conjunction with a fairly common hearing disorder known as *recruitment of loudness*. In a person who has this disorder, the zone between what can just be heard and what is too loud is much narrower than normal. Persons with this affliction would have much difficulty in detecting not only weak sounds but sounds which are fully audible to the normal ear. However, the recruitment ear will retain its sensitivity for loud sound levels. Another problem that a person with recruited ears would face is the discomfort of hearing aids. The hearing aid is a microphone that transmits sounds from the surrounding environment to an amplifier connected to a small loudspeaker built into an earplug and aimed at the eardrum. The major problem is the sounds entering the hearing aid have to be amplified enough to be heard loudly, and at that level the sound may produce discomfort (see Table 1-1).

Table 1-1 Percentage of People with Impaired Hearing in a Noise-Exposed Group

Equivalent Continuous Sound Level (dBA)	Years of Exposure									
	0	5	10	15	20	25	30	35	40	45
80 Risk, %	0	0	0	0	0	0	0	0	0	0
% with impaired hearing	1	2	3	5	7	10	14	21	33	50
85 Risk, %	0	1	3	5	6	7	8	9	10	7
% with impaired hearing	1	3	6	9	13	17	22	30	43	57
90 Risk, %	0	4	10	14	16	16	18	20	21	15
% with impaired hearing	1	6	13	18	22	26	32	41	54	65
95 Risk, %	0	7	17	24	28	29	31	32	29	23
% with impaired hearing	1	9	20	28	34	39	45	53	62	73
100 Risk, %	0	12	29	37	42	43	44	44	41	33
% with impaired hearing	1	14	32	42	48	53	58	65	74	83
105 Risk, %	0	18	42	53	58	60	62	61	54	41
% with impaired hearing	1	20	45	57	64	70	76	82	87	91
110 Risk, %	0	26	55	71	78	78	77	72	62	45
% with impaired hearing	1	28	58	75	84	88	91	93	95	95
115 Risk, %	0	36	71	83	87	84	81	75	64	47
% with impaired hearing	1	38	74	87	93	94	95	96	97	97

Note: Percentage of people with impaired hearing in a nonnoise-exposed group is equal to percentage in a group exposed to continuous sound levels below 80 dBA. Age = 18 years + years of exposure.

Researchers have analyzed noise and its effects on the human ear and have come up with several properties of noise which contribute to the loss of hearing. One of these properties is the *overall sound level of the noise spectrum*. Here it has been noted that noise whose overall sound level is below 80 dB is reasonably safe. However, many industrial noise levels are above 80 dB. To say how these levels would affect the personnel, more specific information on the types of noise would be required.

Another dangerous property of industrial noise is the shape of the noise spectrum. Various studies have revealed that temporary hearing loss, also known as temporary threshold shift (TTS), is a function of the spectrum of the noise. The ear is specifically sensitive to frequencies above 1 kHz, and in fact most cases of hearing loss occur at these frequencies. Noise containing concentrated energy within the octave bands 600—1,200 Hz and higher is much more dangerous to the ear than noise below 600 Hz. Also, a pure tone at a specific level is more dangerous dangerous than a band of noise at the same level, both being at the same frequency.

The next dangerous property of noise is total exposure duration. When exposure to noise permanently lessens a person's sensitivity for hearing, a permanent threshold shift (PTS) has taken place. Through studies of PTS and TTS, it has been determined that the longer a person is exposed to high noise levels, the more his or her hearing ability will decrease. However, the hearing loss due to the noise may not continue until the person is totally deaf.

A final characteristic of noise that should be mentioned is the temporal distribution of noise. However, energy in noise is distributed across time and its final effect on the threshold shift is a function of total energy. It has been determined that partial noise exposures are related closely to the continuous A-weighted noise level by equal energy amounts. The relation between energy and the amount of exposure can be stated as: twice the energy (3 dBA increase) is acceptable for every halving of exposure time, without any increase in danger. Many studies have revealed that significantly high noise levels can be tolerated if exposure time is decreased sufficiently. Table 1-2 indicates permissible noise exposures according to Department of Labor regulations.

Table 1-2 Permissible Noise Exposures

Duration per Day (hours)	Sound Level dBA Slow Response
8	90
6	92
4	95
3	97
2	100
1½	102
1	105
½	110
¼ or less	115

NOISE AFFECTS THE MIND AND OUTPUT

Noise affects the mind and changes emotions and behavior in many ways. Most of the time we are unaware that noise is directly affecting our minds. It interferes with our communication, disturbs our sleep, and arouses our sense of fear. Psychologically, noise stimulates us to a nervous peak. It is overly arousing and presents too high a level of stimulation. Too much arousal causes a person to be too ready and, as a result, he or she will tend to make more mistakes. Experiments in laboratories have revealed that the presence of continuous, loud noises affects the working efficiency in laboratory tasks which require long and concentrated attention. The effects of noise increase the frequency of momentary lapses in efficiency. Noise has its effects on manual workers as well.

Since noise affects a worker's attitude and personality, it also affects his or her output. It can interfere with communication greatly; as the noise gets louder, low speech frequencies become more important. To have 90 percent speech intelligibility between two workers standing about a meter apart, the background noise of mixed frequencies cannot exceed 95 dB, and low-frequency background noise cannot exceed 105 dB (see Figure 1-2).

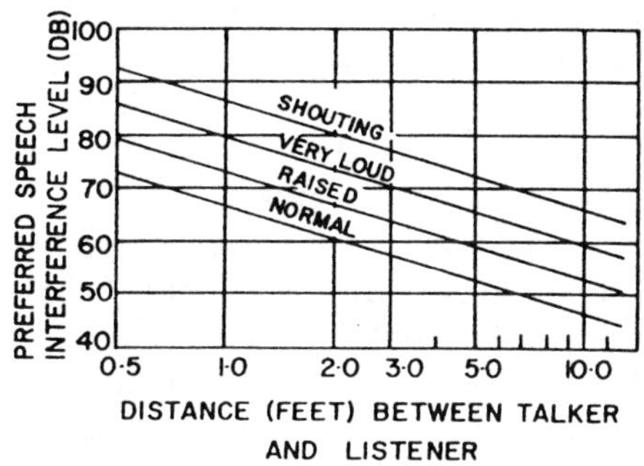

Figure 1-2 Characteristics of speech sound levels in relation to distance.

Noise can cause a decrease in the quality of work output when the background noise exceeds 90 dB. There have been studies, however, which show that the effects of noise on work output largely depend on the type of work. For instance, if a person were put on a time-consuming job that requires constant vigilance, noise would be a degenerative factor in his or her output. High noise levels would tend to cause a higher rate of mistakes and accidents rather than a direct slowdown of production. Results show that a worker's attention to the job at hand will tend to drift as noise levels increase.

EFFECTS ON THE HUMAN BODY

Noise has an explicit effect on the blood vessels, and especially the smaller ones known as precapillaries. Overall, noise makes these blood vessels narrower. It was also found that noise causes significant reductions in the blood supply to various parts of the body. This fact was discovered through a test which employed transillumination, a technique of shining a light inside the mouth of a patient and observing the redness of the cheeks on the outside. If the cheeks were red, sufficient blood was being supplied; if they were pale, there was a deficient amount of blood. When noise was administered around the patient, his or her cheeks became more pale, showing that an insufficient amount of blood was being supplied.

In tests conducted employing a ballistocardiogram used to measure the record of the heart with each beat on a patient in noisy surroundings, it was found that there was a decrease in the stroke volume of the heart. The results of such findings lead to one conclusion—that noise at all levels causes the peripheral blood vessels in the toes, fingers, skin, and abdominal organs to constrict, thereby decreasing the amount of blood normally supplied to these areas. This shrinking of the small blood vessels is known as vasoconstriction and is a reflex action generated by the nervous system. It is triggered by various body chemicals, namely adrenaline, which is produced when the body is under stress. Experiments involving finger pulse amplitude produced similar results. Experiments have revealed that 3 seconds at 87-dB noise constricted arterioles in the fingers and cut down the volume of blood by one half. After the noise stopped, it took approximately 5 minutes for the arterioles to fully recover.

Effects of noise on the blood vessels which feed the brain that were carried out employing a reograph to measure the flow of blood in the vessels leading to the brain concluded that these vessels dilate in the presence of noise. This is the reason why headaches result from listening to persistent noise. Other studies have shown that noise can induce heart attacks. When the small blood vessels are subjected to noise, there is an aggregation of red blood cells within them and the vessels contract in spasm. Thus, noise will actually cause the blood to thicken and clot and could very well lead to a heart attack.

One final part of the body affected by noise is the nervous system. Noise wears down the nervous system, breaks down our natural resistance to disease and our natural recovery, thus lowering the quality of general health.

AUDIOGRAMS

An important step toward hearing conservation in industry is the monitoring of the hearing ability of employees exposed to noise conditions. The monitoring of hearing is known as audiometry. Through audiometry it is possible to detect the least intense sound that can be heard (absolute threshold) and the minimal noticeable difference between two sounds (differential threshold) by an employee. An audiogram is the record of an employee's hearing sensitivity for absolute and differential thresholds as a number of pure-tone frequencies. Hearing sensitivity is measured in terms of the deviation in dB found in normal hearing.

An audiogram is obtained through an audiometer. This device allows pure tones at specific frequencies and intensities to be tested on the employee. There are many types of audiometers used by industry. They can provide frequencies of 500, 1,000, 2,000, 3,000, 4,000, and 6,000 Hz, with output levels more than 70 dB above the standard threshold level.

EFFECTS OF VIBRATION ON MAN

Vibration, like noise, also has harmful effects on the human body. We experience the discomfort of vibration while traveling in boats, cars, and planes.

Many types of tools and machinery with which we have physical contact vibrate and affect our efficiency in using them. Several physical problems directly related to vibration are listed in Table 1-3.

Table 1-3 Vibration-Related Problems.

Effect on Man	Other
Injury	Excessive wear
Fatigue	Excessive noise
Annoyance	Inadequate performance
Interference with performance	Failure to satisfy vibration specifications
Mechanical Failure	
Excessive stress	
Fatigue	
Destructive impacts	

Effects of vibration reveal that sensations due to it are not centralized, since it can be felt throughout the entire body. Figure 1-3 illustrates the subjective response of the human body to vibration as a function of frequency.

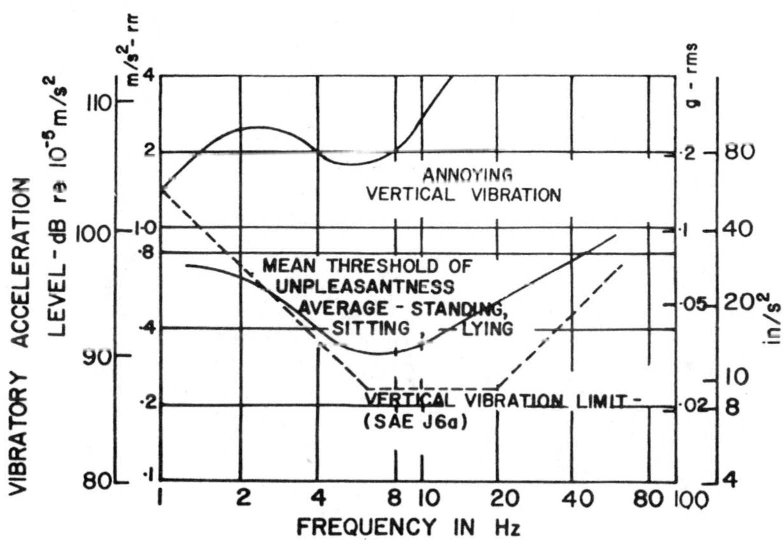

Figure 1-3 Response of the human body to vibratory motion as a function of frequency.

There is a great sensitivity to vibration which occurs in the frequency range of 4 to 10 Hz. Within this range the vibration causes resonance effects of the internal organs and their supports, the upper torso, and the shoulder girdle structures. Many other resonance effects in the body occur also, but these depend on the mode of excitation and the area to which the vibration is applied.

Industrial workers who use certain types of hand-held power tools for extended periods may exhibit Raynaud's phenomenon, whereby the fingers become white and numb as when a person is chilled.

2 *NOISE REGULATION*

OCCUPATIONAL SAFETY AND HEALTH ACT (OSHA)

The Occupational Safety and Health Act was signed on December 29, 1970 and went into effect April 28, 1971. The purpose of this act is to assure so far as possible every working man and woman in the nation safe and healthful working conditions and to preserve our human resources. This act applies to every employer in the country associated with a business that affects commerce. OSHA does not apply to such working conditions that are protected by other federal occupational safety and health laws such as the Federal Coal Mine Health and Safety Act, Atomic Energy Act, Metal and Nonmetallic Mines Safety and Health Standards, and Open Pit and Quarries Safety and Health Standards. This act puts all state and federal occupational safety and health enforcement programs under federal control with the goal of establishing more uniform standards, regulations, and codes with stricter enforcement.

Several of the major aspects of the act maintain federal supervision of state programs to obtain more uniform state inspection under federal standards. OSHA also makes it mandatory for employers to keep accurate records of employee exposures to harmful agents which are required by safety and health standards to be monitored and measured. The law provides procedures to investigate violations by delivering citations and monetary penalties upon the request of an employee. OSHA established the National Institute of Occupational Safety and Health (NIOSH) whose members have the same powers of inspection as members of OSHA. The act also delegates to the Secretary of Labor the power to issue safety and health regulations and standards enforceable by law. This last provision is carried out through the Occupational Safety and Health Administration. OSHA has made many major industries coordinate and revise industrial hygiene, medical, and safety activities with all management functions. With this law in effect, companies have to fully understand their responsibilities in maintaining safe working conditions and the rights of employees. Management decisions on process design, equipment selection, job description, and material use must be made in consideration of the restrictive standards. In the field of employer-employee relations, employers receive demands such as:

- Hazard pay for hazardous jobs and for working time lost due to hazards.
- A union examiner to make tests and evaluate the conditions of the working environment.
- Complete current data on exposure to noise levels, air contaminants, and radiation. Also, industrial hygiene test data on the various levels of exposure.
- More thorough decisions on the size of crews and the length of break periods.
- Complaints of hazardous jobs based primarily on visual observations and sensory response rather than actual data taken from a test situation.
- Physical examinations of employees by a physician of their own choice.

All the preceding demands and complaints must be supported by valid test data obtained from periodic monitoring of the work areas by authorized personnel such as industrial hygienists from headquarters or environmental health or plant personnel directed by corporate technical officials.

Responsibilities of Employers

OSHA enforces two basic duties which must be carried out by employers: first, to provide each employee with a working environment free of recognized hazards that cause or have the potential to cause physical harm or death; second, to fully comply with the Occupational Safety and Health Standards under the act.

To carry out the first duty, employers must have proper instrumentation for the evaluation of test data by an expert in the area of industrial hygiene. This instrumentation must be obtained because the presence of health hazards cannot be evaluated by visual inspection. This duty can be used by the employees to allege a hazardous working situation without any requirement of expert judgment. Evaluation of case histories has revealed that the majority of workers' complaints to OSHA relate to health matters. This duty also provides the employer with substantial evidence to disprove invalid complaints.

Responsibilities of Employees

It states within the act that each employee shall comply with Occupational Safety and Health Standards and all rules, regulations, and orders issued pursuant to this act which is applicable to his own actions and conduct. This law gives employers the right to take full disciplinary action against those employees who violate safe practices in working methods. An employer can also be cited, however, if a compliance officer observes violation of the act by an employee instituting unsafe practices.

Administration

The Secretary of Labor controls the full administration and enforcement of OSHA and has set up a separate agency called the Occupational Safety and Health Administration (OSHA) and the Occupational Safety and Health Review Commission (OSHRC). The act also sets up the National Institute of Occupational Safety and Health (NIOSH). NIOSH is an agency which undertakes research programs to obtain data by which standards are revised. OSHA sets the standards and acts as an enforcement agency, and OSHRC is an appeals agency.

NIOSH has the function of undertaking research and educational functions assigned to the Secretary of Health, Education, and Welfare. NIOSH also develops recommended occupational safety and health standards. The organization of NIOSH is illustrated in Figure 2-1.

The Occupational Safety and Health Administration (OSHA) has full responsibility for administering the Occupational Safety and Health Act. This administration has the full authority to enter and inspect the plant of any employer, to evaluate all working conditions, and to issue any citations or penalties for violations of the standards. An Assistant Secretary of Labor is the administrative officer. Figure 2-2 illustrates the organizational structure of OSHA with all of its offices.

OSHA Standards

There are several different types of standards under this act, one of which is known as a permanent standard. The proposed standard is published in the Federal Register so that people have an opportunity to review it and send in written comments and data on it. Any person may file objections to standards and may also call for a public hearing. OSHA has the authority to officially publish permanent standards and revoke or revise earlier standards on its own basis or on the basis of information submitted by an employer or employee organization, a nationally recognized standards-producing organization, NIOSH, the Department of Health, Education, and Welfare, and a state or political subdivision. Recommendations from such groups must be made within 270 days.

Basically, there are four distinct categories under which OSHA standards fall.

1. Vertical Standards—those that apply to particular industries with specifications that apply to certain operations.
2. Horizontal Standards—standards that apply to all workplaces and relate to very wide areas such as sanitation and illumination regulations.

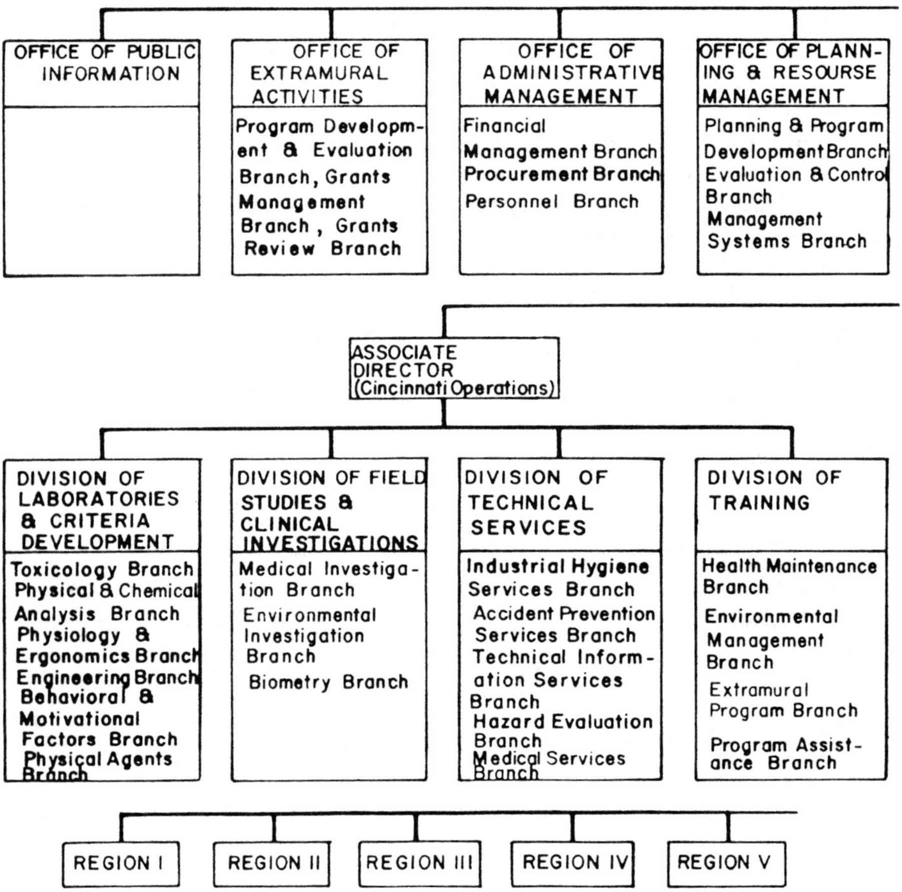

Figure 2-1 Organization of National Institute for Occupational Safety and Health.

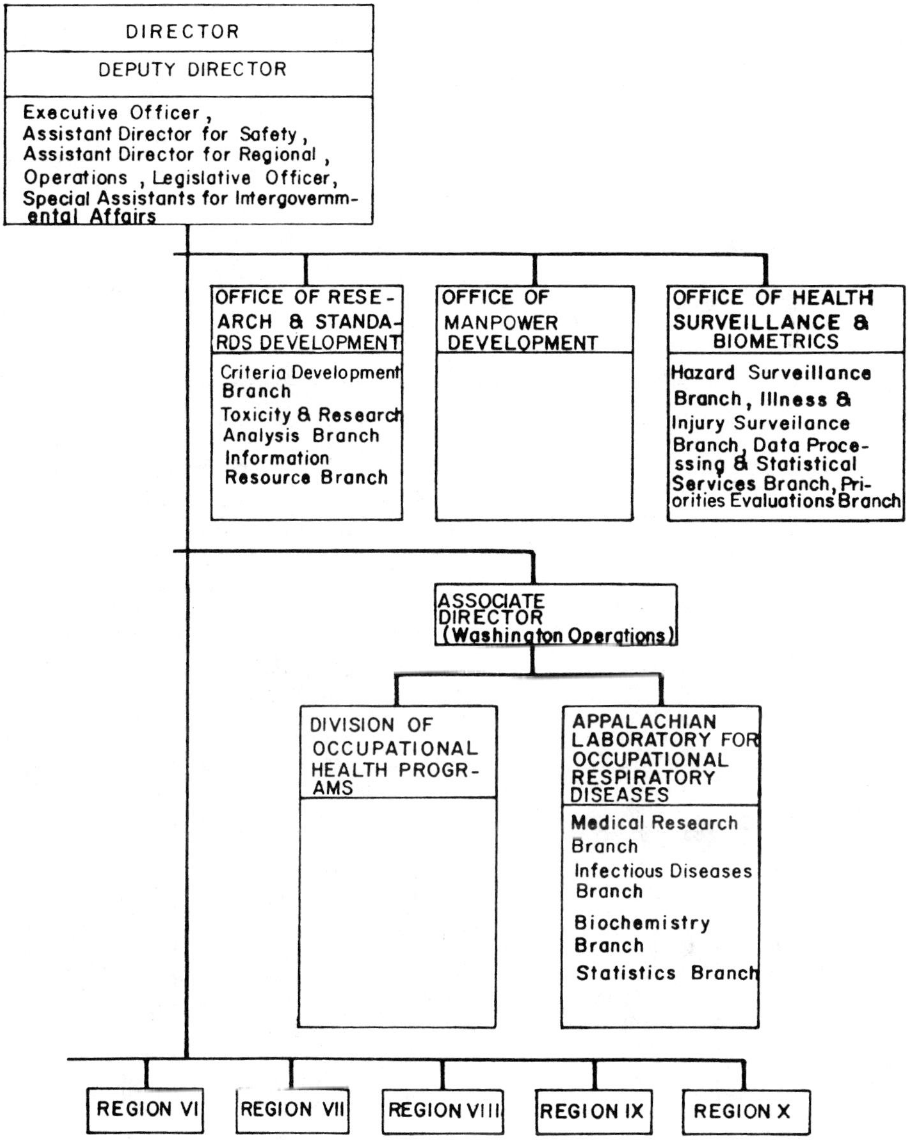

Figure 2-1 (continued).

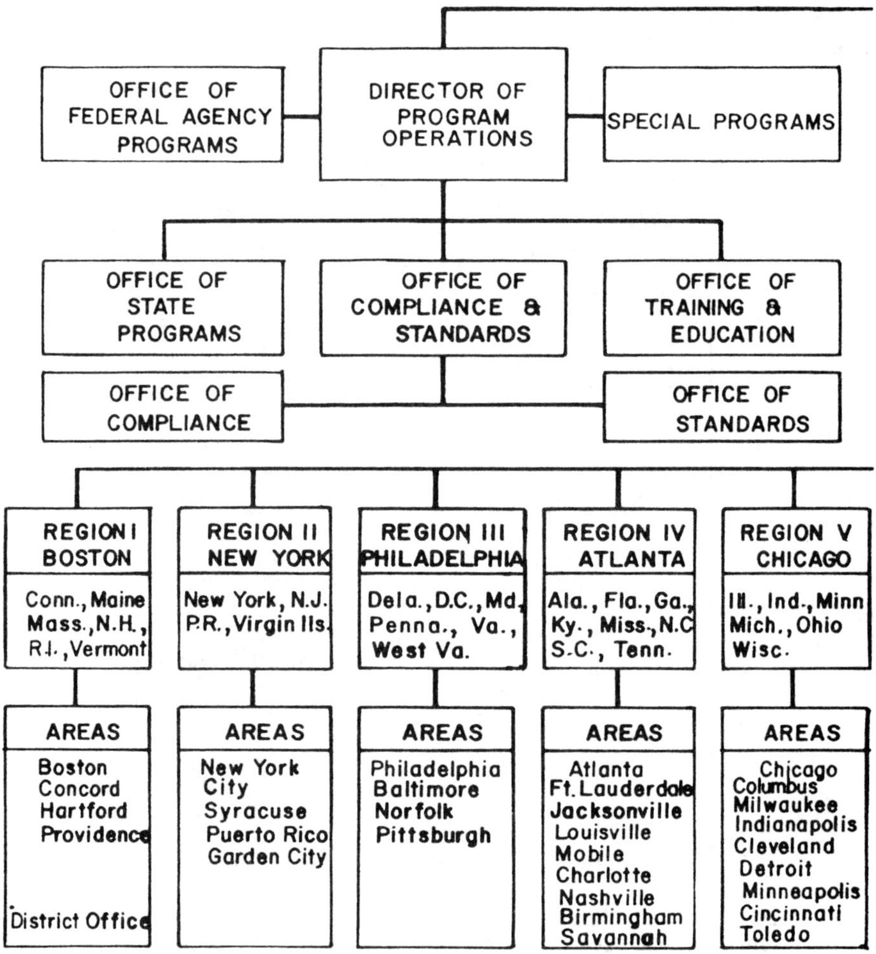

Figure 2-2 Organization of Occupational Safety and Health.

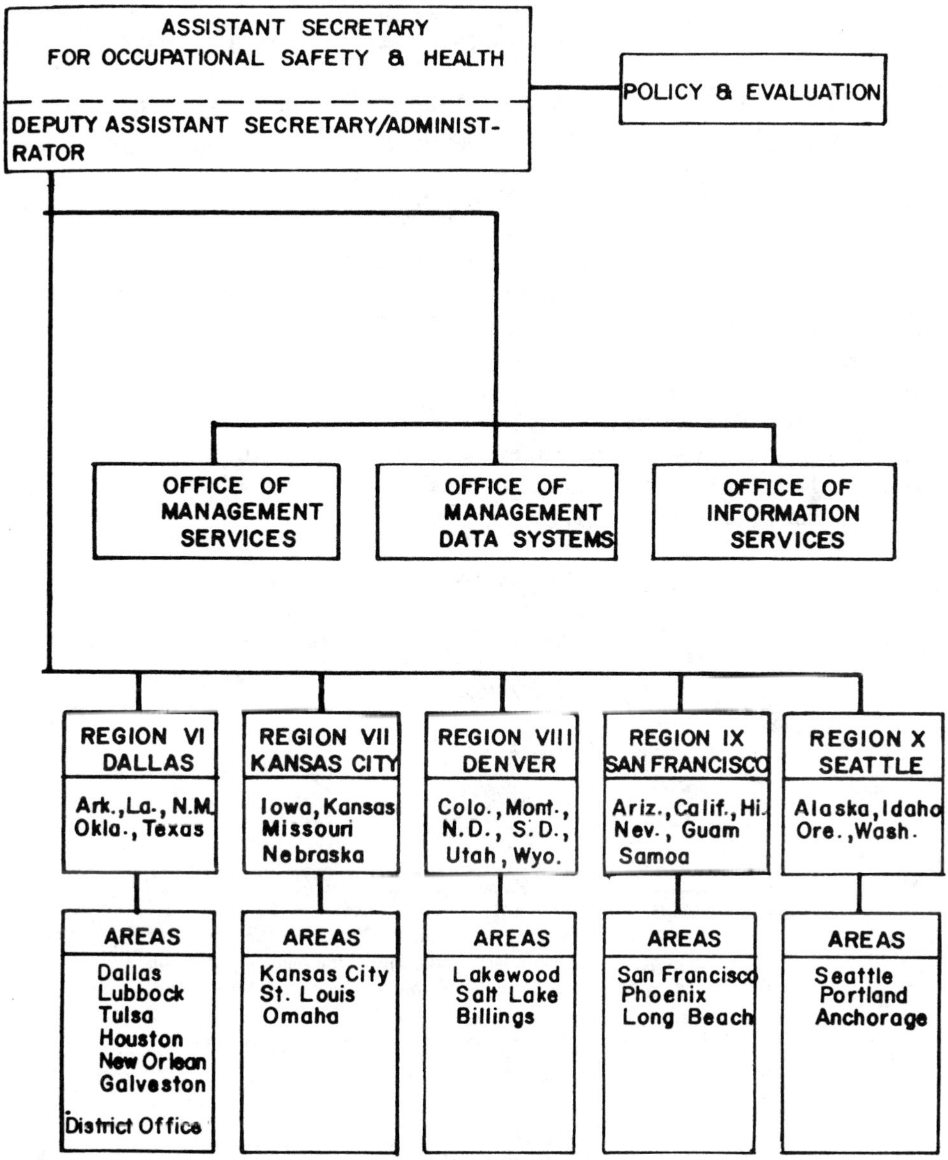

Figure 2-2 (continued).

3. Design Standards—those that provide an OSHA inspector with criteria to cite an employer for violations in the production system.
4. Performance Standards—threshold limit values (TLVs) established by the American Conference of Governmental Industrial Hygienists.

Essentially, a performance standard states the level to which a plant must raise its safety conditions and leaves the method up to the employer.

OSHA'S REGULATIONS ON NOISE CONTROL

Section 50-204.10 of the act establishes acceptable noise levels and exposures for safe working conditions, and gives various means of actions which must be taken if these levels are exceeded. Below, paragraph (a) of this section is quoted:

> Protection against the effects of noise exposure shall be provided when the sound levels exceed those shown in Table I of this Section when measured on the A scale of a standard sound level meter at slow response....

Table 2-1 Permissible Noise Exposures

Duration per Day (hours)	Sound Level dBA Slow
8	90
6	92
4	95
3	97
2	100
1½	102
1	105
½	110
¼ or less	115

[a]Note: When the daily noise exposure is composed of two or more periods of noise exposure at different levels, their combined effect should be considered, rather than the individual effect of each. If the sum of the following fractions: $C^1/T^1 + C^2/TT^2,,, C^n/T^n$ exceeds unity, then the mixed exposure should be considered to exceed the limit value. C^n indicates the total time of exposure at a specified noise level, and T^n indicates the total time of exposure permitted at that level.

Table I of Section 50-204.10 is illustrated in Table 2-1. As can be observed in the table, the acceptable sound intensity for an 8-hour working day is 90 dBA. At a 90-dBA level of exposure for 8 hours the amount of sound energy absorbed is taken as the limit of exposure which will not cause any type of hearing loss in more than 20 percent of those exposed. Table 2-1 also gives higher sound levels with corresponding permissible exposure times that will not induce any more damaging effects than 8 hours of 90-dBA noise. Workers in any industry must not be exposed to sound levels greater than 115 dBA for any amount of time.

Paragraph (a) of this section states that noise levels must be measured on the A scale of a standard sound level meter at slow response. The sound level meter is a measuring device which indicates sound intensity. The A scale on this instrument is one of many, and when operating on this scale the meter functions in a manner very similar to the human ear; it is less responsive to low pitched tones. *Slow response* is a particular setting on the meter, and when the meter is at this setting it will average out high-level noises of short-lived duration.

Impulse or Impact Noise

A quote from paragraph (d) of Section 50-204.10 of OSHA states, "Exposure to impulsive or impact noise should not exceed 140 dB peak sound pressure level." This establishes the maximum sound level to which an employee is to be exposed, regardless of how minute the time span. In Table 2-1 the maximum permissible level for steady noise is 115 dBA. For impact noise a higher level of 140 dB is acceptable because the noise impulse due to impacts (such as explosions) is over before the human ear has time to fully react to it. Levels of impact noise are measured by impact meters or oscilloscopes.

Variable Noise

Regulations of variable noise levels are covered under paragraph (c) of Section 50-204.10 in OSHA. This paragraph states: "If the variations in noise levels in value maxima at intervals of 1 second or less, it is considered to be continuous." Therefore, when the level on the meter goes from a relatively steady reading to a higher reading, at intervals of 1 second or less, the higher reading is taken as the continuous sound level.

Sounds of short duration occurring at intervals greater than 1 second should be measured in intensity and duration over the total workday. These data should be entered into the equation $C^1/T^1 + C^2/T^2 \ldots Cu/Tr$ to obtain the allowable limit. Such sounds may be analyzed using a sound level meter and should not be treated as impact sounds.

Engineering Control Measures

Control measures are covered in paragraph (Cb) of Section 50-204.10 which states: "When employees are subjected to sounds exceeding those listed in Table I of this Section, feasible administrative or engineering controls shall be utilized...." Following are several samples of engineering control measures that reduce noise levels at the source or in the hearing area of the employees.

A. Substitution of Machines

1. Larger, slower machines for smaller, faster ones
2. Step dies for single operation dies
3. Presses for hammers

4. Rotating shears for square shears
5. Hydraulic for mechanical presses
6. Belt drives for gears

B. Maintenance of Equipment
1. Replacement or adjustment of worn and loose or unbalanced parts of machines
2. Lubrication of machine parts and use of cutting oils
3. Properly shaped and sharpened cutting tools

C. Substitution of Processes and Techniques
1. Compression for impact riveting
2. Welding for riveting
3. Hot for cold working
4. Pressing for rolling or forging

D. Dampening Vibration in Equipment
1. Increase mass
2. Increase stiffness
3. Use rubber or plastic bumpers or cushions (pads)
4. Change size to change resonance frequency

E. Reducing Sound Transmission through Solid Materials
1. Flexible mounts
2. Flexible sections in pipe runs
3. Flexible shaft couplings
4. Fabric sections in ducts
5. Resilient flooring

F. Include Noise Level Specifications when Ordering New Equipment

G. Reducing Sound Produced by Fluid Flow
1. Intake and exhaust mufflers
2. Fan blades designed to reduce turbulence
3. Large, low-speed fans for smaller, high-speed fans

H. Isolating Operator
1. Provide a relatively soundproof booth for the operator or attendant of one or more machines

I. Isolating Noise Sources
1. Completely enclose individual machines
2. Use baffles
3. Confine high-noise machines to insulated rooms

The preceding various controls can be achieved with little expense or effort to plant personnel. The Department of Labor expects employers to investigate the possibilities of controlling noise through engineering methods.

Administrative Controls

Sometimes engineering controls are not enough to maintain permissible noise levels. In such cases, administrative controls should be employed to help combat the problem. Several measures which could be taken by the administration are follow.

1. See that workers who have reached the maximum limit of duration for a high noise level, in accordance with Table 2-1, spend the remainder of the day working in an environment with a noise level well below 90 dBA.
2. Arrange work schedules so that employees working most of the day at the 90-dBA limit are not exposed to any higher noise levels.
3. Where the man-hours required for a job exceed the permissible time for one person in a day for the prevailing sound level, divide the work among as many persons as needed, either successively or together, to keep individual noise exposure within the permissible time limits.
4. Perform necessary high-level noise-producing operations during the night or at other times when there is a minimum number of employees exposed.
5. If less than full-time production of a high-noise level piece of equipment is needed, arrange to run it at a portion of each day, rather than all day, for only a few days of the week.

WALSH-HEALEY PUBLIC CONTRACTS ACT

The Federal Walsh-Healey Public Contracts Act regulations took effect on May 20, 1969. To comply with Walsh-Healey regulations on industrial noise exposure, an industry must measure the noise level of its working environment. This type of survey provides valuable data an inspector can use to evaluate working conditions. The data may be obtained by sound survey meters with A-, B-, and C-weighted filters. To test for compliance only the A-weighted measurements are needed. The C-weighted filter has an unweighted flat response between 40 and 8,000 Hz, and can provide additional measurements for a more complete survey. Therefore, any measurement will weigh all the frequencies equally in that range.

Table 2-1 lists the acceptable set of levels under the Walsh-Healey act. Also, the band frequencies are the same for OSHA and Walsh-Healey. When industrial levels exceed permissible limits, the Walsh-Healey act states that a hearing conservation program is mandatory. Steps taken to institute such a program are follow.

1. Group the processes according to noise output; then group the noisiest processes together. This will minimize the number of people exposed.
2. Be selective when purchasing machinery and specify maximum allowable noise output. If no better method is available, cite the Walsh-Healey limits and require the vendor to meet them.

3. Inform maintenance personnel that noise is a by-product of worn or maladjusted equipment.
4. Inspect the operation for noisy machinery to determine which units can operate as well in a noiseproof enclosure.
5. If noise levels still exceed the regulation, require personnel to use ear protection devices.

The U.S. Department of Labor will cooperate with the employers, whenever and however appropriate, to assure the cooperation of employees. Legislation applies to plants with governmental contracts totaling $10,000. In multiplant companies, an inspector may not inspect plants which have no connection with government contracts; but if any one plant is found in violation, the entire company may be penalized. Maximum penalty for a convicted violation is removal from the federal government's approved bidder list for three years. It should be emphasized again that interested parties should obtain interpretation of their own particular case from legal counsel or from the Department of Labor.

PRIMER ON OSHA

A general guide to the Occupational Safety and Health Act of 1970 is herein provided.

Purpose of the Act

All employees of the nation are to be assured healthful and safe working conditions. To accomplish this purpose, mandatory occupational health and safety standards are promulgated. There are also provisions for research, information, education, and training in occupational safety and health. The training and education portions to date have consisted of seminars to familiarize industry with OSHA.

Businesses Affected

The law applies to every employer. This law also applies to construction firms unless the firm is working under the Construction Safety Act due to government contract freelancing. Standards under the Construction Safety Act have been included as part of OSHA standards. An Executive Order by the President has directed each federal agency to draw up safety programs that will in effect be in tune with OSHA. State and municipal employees are excluded from coverage. The act does not apply to working conditions protected under other federal occupational safety and health laws—such as the Federal Coal Mine Health and Safety Act and Atomic Energy Act of 1954.

Duties of Employers

Each employer shall furnish each employee a place of employment free from recognized hazards causing or likely to cause death or serious physical harm. Other duties are as follows:

1. Initiate such programs as may be necessary to comply with the act.
2. Permit only those employees qualified by training and experience to operate equipment and machinery.
3. Provide and train employees in the use of appropriate personal protective equipment where necessary.
4. Make all operating departments aware of requirements under the law. Purchasing departments must require that all equipment and materials meet or exceed recognized good safe practice—the minimum of which is that applicable under OSHA standards. Plant engineers, architectural departments, and construction and industrial engineers should become familiar with good safe practice, OSHA standards, and any consensus standards such as ANSI, ASME, or NFPA standards.
5. Maintain a liaison with the OSHA office in your region and safety officials of your state government.
6. Set specific safety goals for your organization and measure progress toward such goals on a regular basis.

Duties and Rights of Employees

Each employee shall comply with the safety and health standards and all rules, regulations, and orders issued pursuant to the act which are applicable to his or her own actions and conduct. Employees have the right to:

1. Request an inspection if they believe an imminent danger exists or a violation of a standard threatens physical harm.
2. Have a representative accompany a compliance officer during the inspection of any workplace.
3. Advise a compliance officer of any violation of the act which is believed to exist in the workplace, and to question and be questioned privately by a compliance officer.
4. Have regulations posted to inform them of the protection afforded by the act.
5. Have locations monitored in order to measure exposure to toxic materials, have access to the records of such monitoring or measuring, and have a record of their own personal exposure.
6. Be advised of any order issued by the Secretary of Labor granting a variance to an employer, be informed when an application for variance is requested, and petition the Secretary for a hearing pertaining to any variance requested.

7. Request the employer to review copies of regulations and standards.
8. Be given medical examinations or other tests to determine whether their health is being affected by an exposure, and have the results of examinations or tests furnished to their physicians.
9. Be informed of citations made to the employer through the requirement that citations be promptly posted, contest the length of time fixed for abatement of the violation on the basis that it is unreasonable, and participate in hearings concerning abatement time.

It is suggested that all levels of management be informed of the duties and rights granted to employees under the Occupational Safety and Health Act.

Occupational Safety and Health Standards

Employers are obligated to familiarize themselves with those standards which apply to them. It should be pointed out that OSHA standards are minimum requirements. Employers must be familiar with or have access to any of the consensus standards that may apply to their operation.

Interim Standards. The National Consensus Standards (ANSI and NFPA) are those on which interested parties have reached substantial agreement after considering diverse views—plus existing federal standards.

Emergency Standards. The Secretary of Labor determines the standards necessary to protect employees exposed to toxic substances from physical harm or from new hazards. Once published, the Secretary must follow by setting permanent standards.

Permanent Standards. These standards are set by the Secretary of Labor and most adequately assure, to the extent feasible and on the basis of the best available evidence, that no employee will suffer material impairment of health or functional capacity even if regularly exposed for his or her working life.

Administration

Administration and enforcement are vested in the Secretary of Labor and the Occupational Safety and Health Review Commission. Research is vested in the Secretary of Health, Education, and Welfare and will be carried out by the National Institute for Occupational Safety and Health.

Complaints of Violations

Any employee who believes that a violation of a standard or any provision of the act or any unsafe or unhealthy condition exists may request an inspection by (1) a signed written notice to the Department of Labor, or (2) by an outline of the grounds for notice. A copy of the employee's request must be provided to the employer no later than the time of inspection. The copy need not include the employee's name. If the Secretary of Labor finds no reasonable grounds and no citation is issued, complainants must be notified in writing of the final disposition. The Secretary is required to set up procedures for informal review where a citation is not issued.

Inspections

Compliance officers may enter upon presenting appropriate credentials to the owner, operator, or agent in charge. They are authorized to enter without delay, at any reasonable time, to inspect and investigate during regular working hours. They may question privately an employer, owner, operator, agent, or employee. The employer and employee representatives may accompany the compliance officer to aid in the inspection. Any person giving advance notice of impending inspection is liable for a $1,000 fine, six months' imprisonment, or both. It is a federal crime to kill, assault, or resist personnel carrying out their duties. Trade secrets are to be treated confidentially.

Upon completion of an inspection, the compliance officer is to confer with the employer to advise of violations found. A citation will not be issued at this time. It will be issued by the area director via registered mail. Reasonable time is fixed for abatement. Each citation must be posted at or near the place of violation. Time for abatement is listed on serious and nonserious violations. No citation can be issued after the expiration of six months following the occurrence of that violation.

Notification of Proposed Penalty

Employers have 15 working days to notify the Department of Labor they wish to contest a citation or the penalty, or request a variance. If they fail, the citation and penalty are final. Upon review, the review commission will affirm, modify, or vacate the citation or penalty. Review of commission orders can be appealed to the U.S. Court of Appeals. If the penalty is contested, it is not final until reviewed and affirmed by the commission.

Contesting a Citation

Notification must be made to the area director in charge of the local office of the U.S. Department of Labor which initiated the action being contested. The time allowance is 15 working days from receipt of a letter by certified mail containing the Department's proposed penalty. This notification is the *Notice of Contest* and when placed in the mail, it becomes Act I-Scene I of the case which will thereafter go before the Occupational Safety and Health Review Commission. If any of the employees who work at the site of the alleged violation are unionized, a copy of the Notice of Contest must be served upon their union. If any of the employees who work at the site of the alleged violation are not represented by a union, copies of the Notice of Contest must either be posted in a place where employees will see it (the Labor Department's regulation for posting of citations) or be personally served upon such employees. There is no particular form prescribed for the Notice of Contest, but it must clearly identify the citation, notification of proposed penalty, or notification of failure to correct the violation which forms the basis for its filing. A Notice of Contest must also contain a listing of the names and addresses of those parties to the case who are caused to be personally served with the same, plus the address of the posting when one is required.

Time for Abatement of Hazards

A citation prescribes a reasonable time for compliance. The time limit may also be contested within 15 working days. The time limit does not begin until there is a final order of the review commission.

Employees also may object within 15 working days to the time fixed in the citation on the grounds of alleged unreasonableness, but only if the employer does not contest. The same review procedures apply as before.

Variances

Employers can ask the Secretary of Labor for a variance order if they cannot comply with a standard prescribed by the law. There are three types of variance:

1. *Temporary*—when an employer needs more personnel or larger facilities in order to comply with the standard.
2. *Equally Safe Working Conditions*—when an employer can provide just as safe and healthful working conditions as if he or she complied with the standard.
3. *Safety and Health Experiment*—when an employer takes part in a worker safety or health experiment with approval from the Secretary of Health, Education, and Welfare.

Failure to Correct Violations Within Allowed Time

If an employer fails, the Sécretary will notify the employer by certified mail of this failure and the assessed penalties. This notification will be final unless contested within 15 days. In good faith—if the employer cannot comply due to factors beyond his or her control—a hearing will be afforded to affirm or modify the time requirement.

Penalties for Violations

Penalties for violations are as follows: those not corrected in time—up to $1,000 per day per violation; for willful or repeated violations—up to $10,000 per violation; for willful violations resulting in death—$10,000 or six months' imprisonment (a second conviction doubles this penalty).

Recordkeeping Requirements

OSHA 100—mandatory; OSHA 101—optional, other reports may be used if they contain all information required; OSHA 102—mandatory.

Statistics

The Secretary of Labor and Secretary of Health, Education, and Welfare are required to develop and maintain effective programs for collection, compilation, and analysis on work injuries and illnesses.

Imminent Dangers

In the event of conditions or practices, so determined, which might cause death or serious physical harm, a court order can be sought to restrain normal operations by shutdown of operations or plant, or by removal of employees from the affected area.

Protection Against Harassment

No persons shall be discharged or discriminated against because they exercise any right under the act. If so, they may file a complaint within 30 days of the illegal action with the Secretary of Labor. The Secretary of Labor takes the issue to U.S. District Court for appropriate relief.

What an Employer Should be Doing

To assure efficient and orderly organization relating to existing practices and OSHA, a list of priorities to assure compliance with OSHA, while at the same time maintaining an overall program, should be established. Procedures and practices should be reviewed in establishing a plan of action and priorities.

1. Formulate a policy statement outlining duties and responsibilities to be included as a part of the company's program.
2. Make certain appropriate personnel in the organization are knowledgeable about OSHA and are kept up to date on new developments.
3. Establish an OSHA-required recordkeeping system as outlined in the Department of Labor pamphlet entitled *Recordkeeping Requirements under the Williams-Stieger Occupational Safety and Health Act of 1970.*

4. Analyze and, where necessary, strengthen safety and health programs.
5. Establish procedures for handling a citation should one be received.
6. Improve liaison with insurance brokers and/or carrier loss prevention personnel. These sources usually can offer expertise in safety and health matters.
7. Establish hazard control priorities through surveys of premises, equipment, and materials based on applicable safety and health standards.

GUIDE FOR IMPLEMENTATION

Policy

The policy of the manufacturing organization should be to provide safe and healthful working conditions in all activities. Consistent with that policy, all managers within an organization are responsible for compliance with OSHA as an integral part of their safety responsibility.

Procedures

Each manager shall assume the responsibility for applying the provisions and standards of OSHA and shall specifically designate two alternates, in a specific order, to serve as the persons responsible in his or her absence. Under the direction of the manager, there shall be:

1. A specifically established safety program, in writing, which produces results.
2. An active safety committee.
3. A prompt response to recommendations submitted relating to safety or fire prevention.
4. A thorough and effective accident investigation and reporting procedure.
5. A training program for employees and supervisory personnel directly related to possible injury or illness in the location's operations.
6. A specific audit or survey of all premises, equipment, and material so that recommendations can be developed to obtain compliance with established standards.

Recommendations may be categorized as follows:

- Immediate attention
- Imminent dangers
- Those items only requiring modest expenditures of time or money

- As schedule permits—those items which are temporarily acceptable but should be corrected
- Preplanning necessary—recommendations requiring new programs, equipment, or processes

It is recognized that some technical violations which do not constitute hazards may be attended to during major renovation or new planning. Action on all of the preceding conditions should be well documented as evidence of good faith. Immediate attention is required to any complaints on the part of employees concerning a possible injury or illness potential. Also necessary is a communication system through which all personnel are kept abreast of new standards or procedures as they are published by the Department of Labor. Specific goals are to be established for the safety program and progress measured toward those goals on a regular basis.

Should there be an inspection by a Department of Labor representative, the inspector should be immediately referred to those management personnel designated as responsible for compliance with the act, but preferably the location manager. Examine the inspector's credentials to assure that he or she is a representative of the Department of Labor. It is important to recognize that by law Department of Labor inspectors have the right to make an inspection, usually unannounced.

Before the inspection, attempt to review with the inspector company policy concerning the act; safety program organization and procedures; evidence of safety activities such as minutes of safety meetings, reports submitted after plant inspections, educational activities, noise control, and so on; accident data to emphasize the thoroughness of investigations and the procedures for taking corrective action after an accident occurs. Arrange for the availability of employees who have a right to accompany the inspector during his or her inspection. Management personnel should make themselves available for discussions with the inspector during and after the plant tour. An exit interview should be insisted upon and a record made of all that the inspector says, particularly if citations are to involve conditions that would either be difficult to correct or require sizeable expenditure. It is important to understand that there is but a 15-day response period to citations after they are received. Upon receipt of the inspector's report, immediately communicate with appropriate headquarters personnel to establish a plan of action in response to recommendations and possible citations.

Recordkeeping Requirements

Each location must be provided a copy of the U.S. Department of Labor publication *Recordkeeping Requirements under the Williams-Stieger Occupational Safety and Health Act of 1970*. Forms required are: OSHA 100—Log of Occupational Safety and Health Injuries and Illnesses; OSHA 101—Supplementary Record of Occupational Injuries and Illnesses; OSHA

102--Summary, Occupational Injuries and Illnesses. Two of the forms, OSHA 100 and OSHA 102, are mandatory.

First Report of Injury forms completed for workmen's compensation purposes may in some cases be used in place of OSHA 101 provided that information required by OSHA 101 which is not contained in a First Report of Injury is added. OSHA 100 is a continuous log of all recordable injuries. Instructions are contained on the reverse side of the form concerning what is to be recorded.

OSHA 102 is a calendar year summary of data contained in the Log of Occupational Injuries and Illnesses which must be completed by January 31st for the preceding year. One copy should be retained for file purposes. One copy must be posted where employees can examine it. No length of time for posting this form has been established. It is suggested that it be placed in a glass-covered frame and remain posted for 30 days.

Each accident or health hazard that results in one or more fatalities or hospitalization of five or more employees must be reported to the area director of the Occupational Safety and Health Administration within 48 hours.

Each recordable occupational injury or illness must be entered on OSHA 100 within two working days of receiving information that the case has occurred. Samples of each of these three forms are to be attached. All forms are to be kept at the location where employees work for a period of five years.

Forms required by OSHA should be kept in one file or folder and be available for review by a Department of Labor compliance officer. Within the publication *Recordkeeping Requirements under the Williams-Stieger Occupational Safety and Health Act of 1970* is a centerfold poster which must be displayed in a prominent place. Employers have been cited for not having this poster displayed, so it is important that it be done.

NOISE ENFORCEMENT DIVISION
U.S. ENVIRONMENTAL PROTECTION AGENCY

On April 30, 1976, the administrator of the U.S. Environmental Protection Agency established the Noise Enforcement Division under the Deputy Assistant Administrator for Mobile Source and Noise Enforcement, Office of Enforcement. Responsibilities are divided into the following four general enforcement areas:

1. general products noise regulations
2. surface transportation noise regulations
3. noise enforcement testing
4. regional (EPA), state, and local assistance

The general products group develops and helps implement strategies and regulations for the enforcement of noise emission standards for a variety of new products. EPA has also promulgated noise standards and regulations for the control of noise from portable air compressors. Additional regulations have been developed to control noise from truck-mounted solid-waste compactors, truck transport refrigeration units, and wheel and track loaders and dozers.

Federal enforcement strategies are applied to new products to assure their conformance with noise standards and typically include:

- production verification of configurations or classes of products, that is, testing of initial representative products at the assembly line
- selective enforcement auditing of products, that is, statistical sampling and testing of new products at the assembly line
- manufacturer's time-of-sale warranty
- tampering prohibitions
- maintenance instruction requirements
- recall of noncomplying products

State and local jurisdictions may adopt and enforce in-use standards applicable to noise emissions resulting from the operation of interstate rail carriers and interstate motor carriers only if such standards are identical to the federal standards. However, a waiver of this federal preemption may be granted if the administrator of EPA determines that such state or local regulation is necessitated by special local conditions and is not in conflict with federal regulations for interstate rail and motor carriers. Types of state and local noise regulations include the following:

1. product performance standards implemented through licensing or certification procedures
2. operational limitations, such as curfews
3. movement limitations, such as restrictions on truck traffic in noise-sensitive areas, for example, hospital zones
4. property line limitations according to land use, for example, maximum noise emission levels at property lines of residential zones
5. nuisance prohibitions

3 ACOUSTICS

NATURE OF SOUND

Sound waves are pressure variations produced as a result of a mechanical disturbance in a material medium. They can be classified into two main categories according to whether or not that medium is strained beyond its elastic limit. Our attention will be confined to common sounds and vibrations where the medium remains within its normal elastic limits. Shock waves and other impulsive sounds which exceed these physical limits require a different approach with regard both to theory and instrumentation.

In acoustical measurements we generally measure the pressure variations about the local ambient and static pressure of the medium, although it is possible also to measure either the local particle velocity or even particle displacement. It has been found that pressure-sensitive transducers give the most accurate electrical reproduction of the sound waveform.

The production, propagation, and detection of sound waves are generally related to the setting up of an oscillation. The simplest form of oscillation is simple harmonic motion; when pressure variations of this form are produced, the resulting sound is referred to as a pure tone because the pressure variations occur at only one frequency. Sound-wave propagation can be considered in the form of plane or spherical waves. Where the source is effectively a single point, spherical waves result. But at large distances from such a source, where the curvature of the wave front is small, plane waves can be considered to exist. The plane wave effect is similar to what would exist if a large plane surface were to vibrate so that all points on the surface were in phase.

If we now consider the simple harmonic vibration of such a plane surface in air or any other medium, as in Figure 3-1, it is apparent that movement of the surface causes a slight variation in the local pressure close to the surface. This will be transmitted through the medium at a particular velocity of propagation, as are the ripples on the surface of still water. The velocity is proportional to the elasticity and density of the medium. In an ideal gas, the relationship is:

$$V = \sqrt{\frac{\gamma z}{\rho}}$$

where V is the velocity of propagation, z the local atmospheric pressure, ρ the ratio of the principal specific heats of the gas, and γ the local density. From the gas laws, the velocity of propagation in a particular gas will be proportional to the temperature according to the relation:

$$V = \sqrt{\gamma RT}$$

where R is the gas constant and T the absolute temperature. Although we normally think of sound propagation in gases, particularly in air, the basic acoustical principles apply equally well in solids. Table 3-1 gives the velocity of sound propagation in various materials.

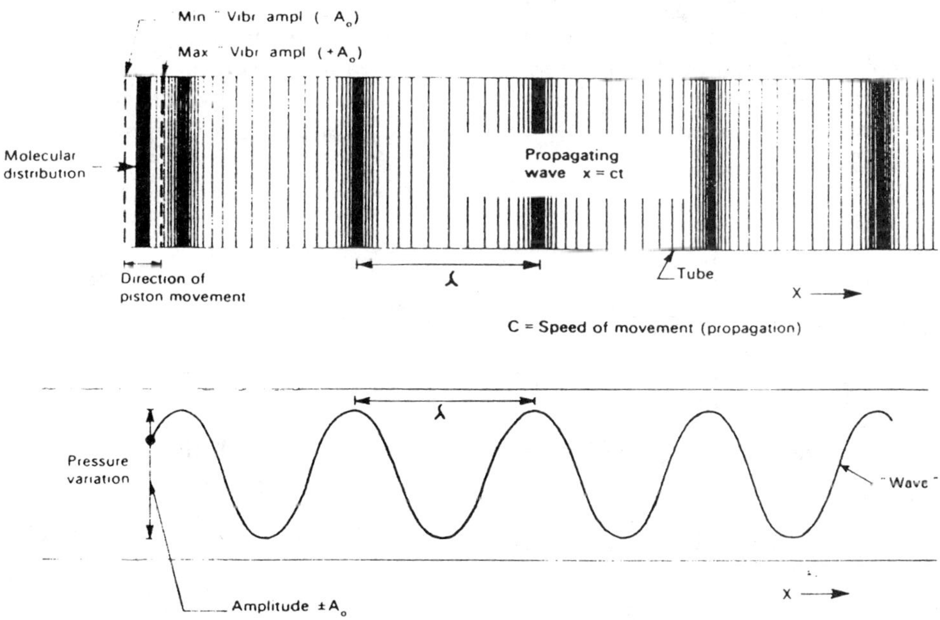

Figure 3-1 Illustration of how sinusoidal vibration of a plane surface are transformed into a sinusoidal pressure wave.

Table 3-1 Velocity of Sound Propagation in Various Media

Medium	Velocity of Propagation (m/sec)
Air at 1 atm 0°C	332
Air at 1 atm 100°C	386
Air at 25 atm 0°C	332
Hydrogen 0°C	1269
Vulcanized rubber	54
Granite	6000
Mild steel	5050
Aluminum	5000
Lead	1200
Water 15°C	1440
Salt water 15°C	1470
Alcohol	1213

If propagation is truly plane wave, then the variation of local pressure at a distance from a source, such as that in Figure 3-1, will also be sinusoidal. The properties of the wave are illustrated in Figures 3-2 and 3-3. Figure 3-2 shows the pressure distribution at any instant in time in the sound field, whereas Figure 3-3 shows the pressure-time variation at any point. While the sound pressure varies through a complete cycle in time T the wave front will move with velocity of propagation V, where $V = \lambda\ T$, or alternatively $V = n\lambda$ where n is the frequency of the vibration, which is defined as the number of periods completed in unit time. When the period is measured in seconds, the frequency is in cycles per second, or Hertz. The relationship between frequency and wavelength is shown in Figure 3-4 for the particular case of sound propagation in air.

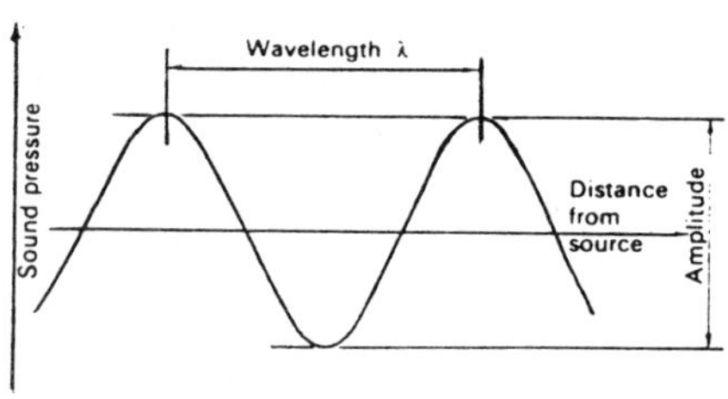

Figure 3-2 Time properties of a simple harmonic oscillation.

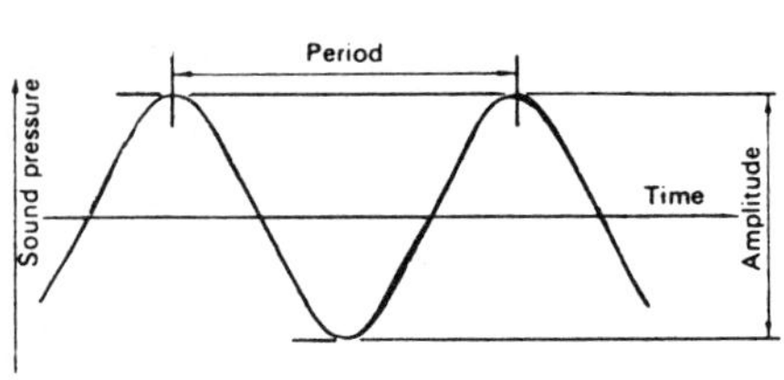

Figure 3-3 Distance properties of a simple harmonic oscillation.

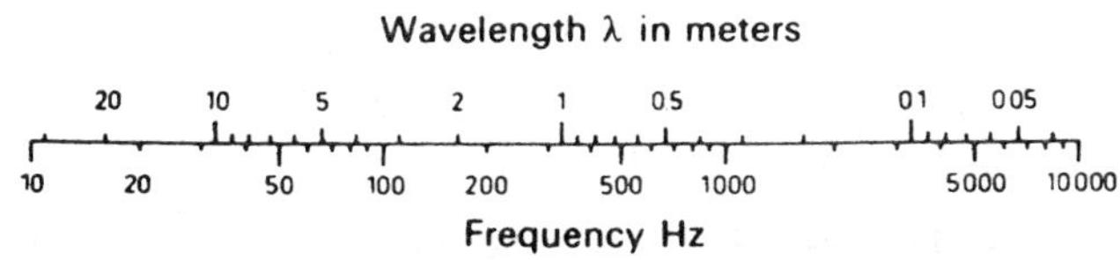

Figure 3-4 Relationship between frequency and wavelength for sound in air at normal atmospheric conditions.

The strength of the pressure wave can be measured by the amplitude of the local pressure fluctuations around the ambient atmospheric pressure. Such pressure variations are an extremely small proportion of the atmospheric pressure, and the most common unit of measurement is the Pascal (Newton per square meter) which is approximately 10^{-5} atmosphere.

THE SOUND FIELD

Although this description has been for a plane, free, progressive wave, similar concepts apply for spherical waves which are encountered in field measurements. The only difference is due to the enlarging area over which the energy of the source is spread as the wave progresses. This gives rise to the concept of intensity, which is the average acoustic power per unit area. A practical illustration of intensity is a sound which is painful when the source is close to the ear but is inaudible at a distance. Assuming that the emitted noise power is W, this spreads out from the source in the form of spheres with continuously increasing radius r (Figure 3-5).

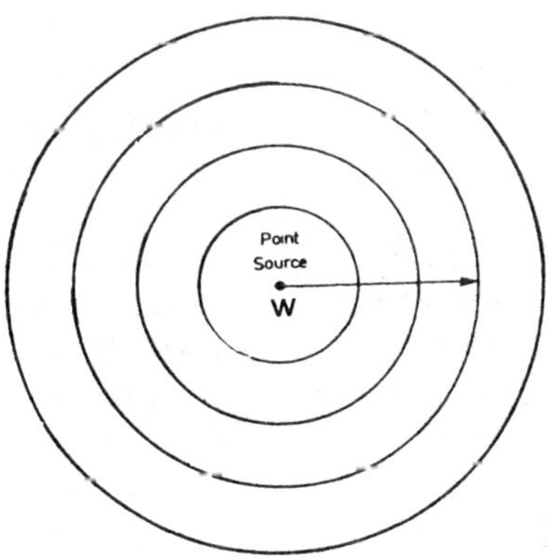

Figure 3-5 The concept of spherical wavefronts.

Thus, the noise power passing through a unit area on the surface of a sphere of radius, r, must in a nondissipative medium equal the emitted power, W, divided by the area of the surface ($A = 4\pi r^2$):

$$I = \frac{W}{4\pi r^2}$$

Now, at a sufficient distance away from the source, this average noise intensity, I, is also proportional to the square of the sound pressure, p, at r, whereby:

$$p = \alpha\frac{1}{r}$$

This relationship is the inverse (distance law—sometimes also called the *inverse-square law*—and governs the free sound radiation in the acoustic far field) of a sound source.

Spherical radiation from sound sources and plane wave fronts at measuring points in space are ideal conditions of sound propagation which are seldom encountered in actual practice. When one or more sound-reflecting objects are present in the field, the wave picture changes completely due to the reflections. This is easily understood as there is not only the original progressive wave present, but also a reflected wave traveling more or less in the opposite direction to the original wave. The sound pressure at a certain point in the field is, at any instant, the combination of the pressure due to the original wave and the pressure(s) due to the reflected wave(s).

Before discussing the effects of reflection, the conditions necessary to obtain a reflection at all should be briefly mentioned. If for instance an object which is very small compared with the sound wavelength is placed in the sound field, no real reflection of the wave is produced. To produce a reflection which effectively interferes with the sound field, the dimensions of the reflecting object in the directions perpendicular to the direction of travel of the original sound wave must be on the order of the sound wavelength, or larger. The amount of reflection also depends on the sound-absorbing properties of the object. Thus, both the physical dimensions and the sound absorption of an obstacle affect its reflecting properties. Returning to the effects of reflections upon the original sound wave, one of the most important effects with regard to noise measurements is the production of a so-called diffuse sound field. A diffuse sound field is a field in which a great number of reflected waves, from all directions, combine in such a way that the average sound energy density is uniform everywhere in the field. An approximation to this kind of field exists in large, reverberant rooms, and the way in which it is achieved is demonstrated in Figure 3-6.

To illustrate the importance of the previously described kinds of sound fields with respect to noise measurements, assume that a noise source is placed in a room. The sound pressure level set up around the item as a function of distance might then be of the type shown in Figure 3-7.

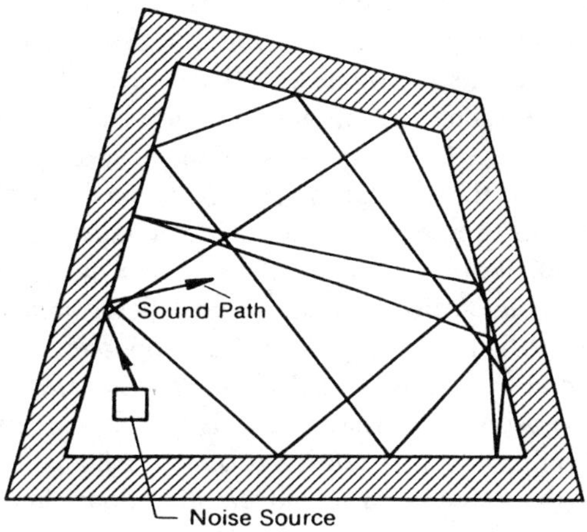

Figure 3-6 Illustration of how sound diffuseness is obtained.

The boundary between the near and far field is determined by the wavelength of the sound wave and by the distance between the sound source and the point of measurement. In the near field, the sound pressure heavily depends on the radiating characteristics of individual noise-generating components, and

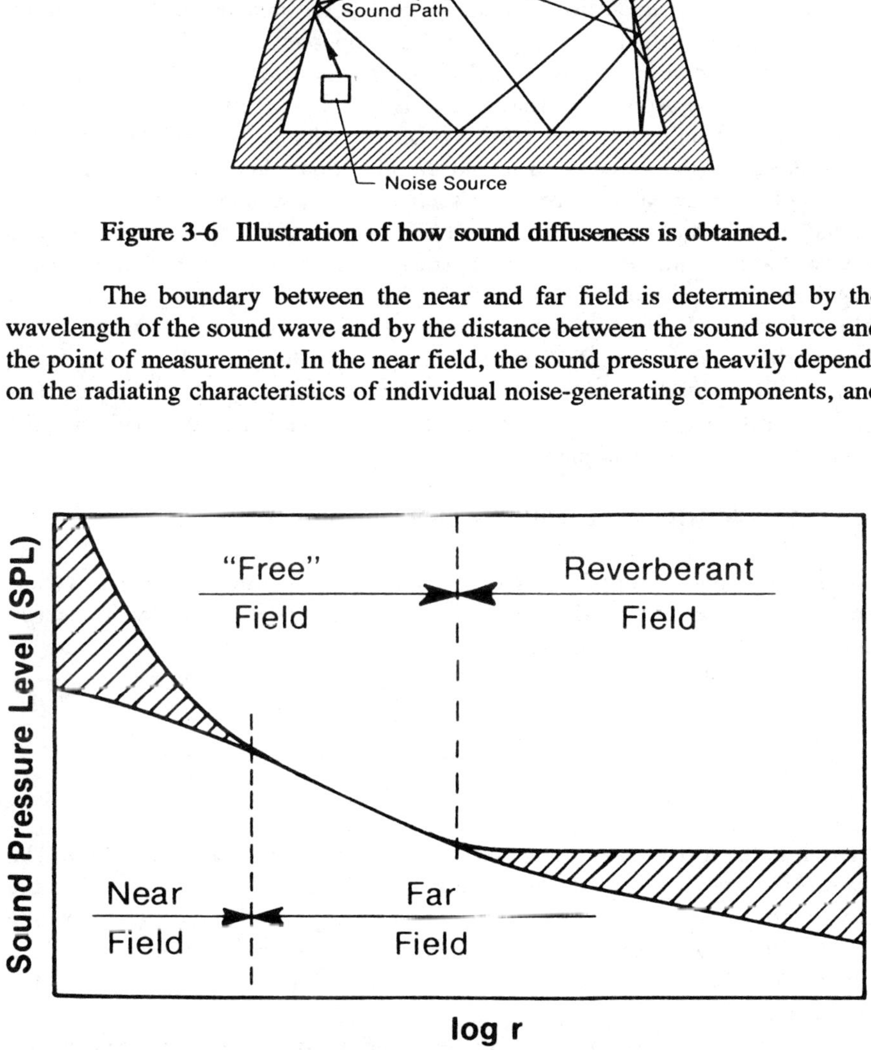

Figure 3-7 Sketch illustrating the variation of sound pressure level in a room as a function of distance from a source. The shaded areas indicate regions of level fluctuating with position.

the particle velocity does not necessarily have the same direction of travel as the sound wave. The result is that small changes in position show large differences in pressure. In the far field, particle velocity is primarily in the direction of the propagation of the sound wave, and pressure is inversely proportional to distance. In practice, the boundary between the near and far fields is taken to be at a distance of about three times the largest dimension of the largest radiating surface of the sound source. In the far field, measurement can be made of the sound actually radiated by the source. And these data can be used to predict sound pressure at greater distances. In the near field, component sources can be measured and identified by using a probe microphone.

The free and reverberant field boundary is determined by the environment in which the acoustical measurement is made. A free field is one in which there are no reflections. This condition is often simulated in anechoic rooms for measurement of the radiated noise and the directional characteristics of sound sources. But in actual environments such as outdoors and factory areas, the ground or floor and walls reflect sound. The result is that more than one wave is present at the point of measurement. The resultant pressure is higher or lower, depending on amplitude and phase differences between the primary wave and the reflected waves. In the free field the directly radiated sound dominates, but in the reverberant field the effect of reflections dominates. Reverberant fields are poor conditions under which to evaluate sound radiated by sound sources. But this has no bearing on environmental noise measurements made in the community or at the workplace.

SCALES FOR MEASUREMENTS—THE DECIBEL

In characterizing the magnitude of a sound pressure wave, the rms (root mean square) value is used because it has a direct relationship to the energy content of the signal. It is defined as:

$$A_{rms} = \sqrt{\frac{1}{T}\int_{o}^{T} a^2(t)dt}$$

where a is the instantaneous amplitude and T is the period of observation. The quantity normally measured when dealing with acoustic noise is therefore the rms sound pressure. Sound pressure is the force per unit area caused by the sound wave, and the unit of measurement is the Newton/m^2, referred to in the International Standards Organization Recommendation 1000 for S.I. units as the Pascal (Pa).

The weakest sound pressure perceived as sound is a very small value, and the range of sound pressure perceived as sound is extremely large. The weakest sound-pressure disturbance to be detected by an "average" person at 1,000 Hz has been found to be 20 μPa (approximately 0.003 μpsi). On the other hand, the largest sound pressure perceived without pain is of the order of 10^7 μPa. Because the scale of sound pressures covers such a wide dynamic range, the use of a linear pressure scale is considered to be impractical. Also, the hearing mechanism responds nonlinearly to changes in sound pressures. It is, therefore, convenient to use a nonlinear scale for sound pressure measurement. Such a scale is the decibel or dB scale.

The decibel is defined as ten times the logarithm to the base ten of the ratio between two quantities of power. As the sound power is related to the square of the sound pressure, convenient scale for sound measurements is defined as:

$$\textit{Sound Pressure Level} = 10 \log (P^2/P_o^2) = 20 \log (P/P_o)\ dB$$

where p is the rms level of the sound pressure being measured and p_o is a reference sound pressure, normally taken to be 20 μPa. It should be noted that the term *level* has been introduced in the preceding equation. This indicates that the given quantity has a certain level above a certain reference quantity. It reduces this dynamic range of 1:10,000,000 in sound pressure to an equivalent 0 to 120 dB, a much more manageable set of numbers.

Shown in Figure 3-8 are some examples of the level of familiar noise sources in our environment. Circumstances often arise where it is necessary to add two or more decibel levels. Of course, it is not possible to make this addition algebraically because of the logarithmic scale. The levels must be combined on an energy basis. The procedure for doing this is to convert the decibel values to power ratios: add or subtract these ratios as required and then reconvert to decibels. A graph which conveniently eliminates logarithmic manipulation is given in Figure 3-9. The graph shows that combining any two equal levels results in a 3-dB increase in total level. It can also be seen that if two levels differ by more than about 10 dB, then the sum of the two levels will be insignificantly different from the higher of the individual levels. From this fact is derived the rule of thumb that the source noise to be measured should be at least 10 dB above background level.

LOUDNESS AND OBJECTIVE MEASUREMENTS

The human ear exhibits a nonlinear frequency response. Moreover, frequency response varies with sound-pressure level. Figure 3-10 shows a set of equal loudness curves for an average young person with good hearing, listening to pure tones in a sound field. The ear is most sensitive around 3,500 Hz with a marked reduction occurring at low frequencies. With increasing level, the response of the ear becomes more flat. Each curve is assigned a Phon value equal to the sound pressure intercept at 1 kHz. For example, each point on the 40 Phon curve is as loud as 40 dB SPL at 1 kHz.

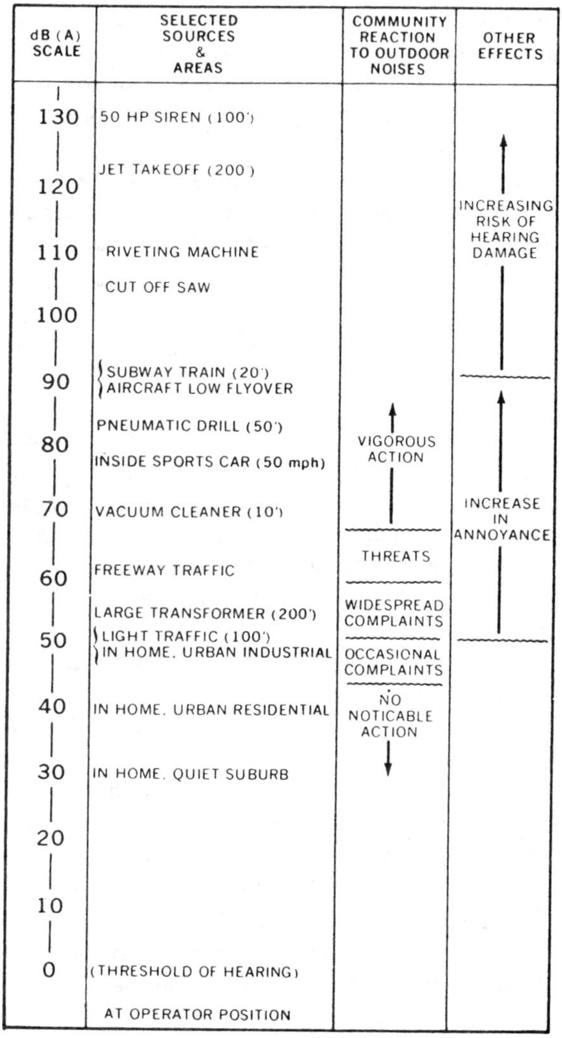

Figure 3-8 Noise levels in dBA, community reaction, and other effects for some typical sound sources.

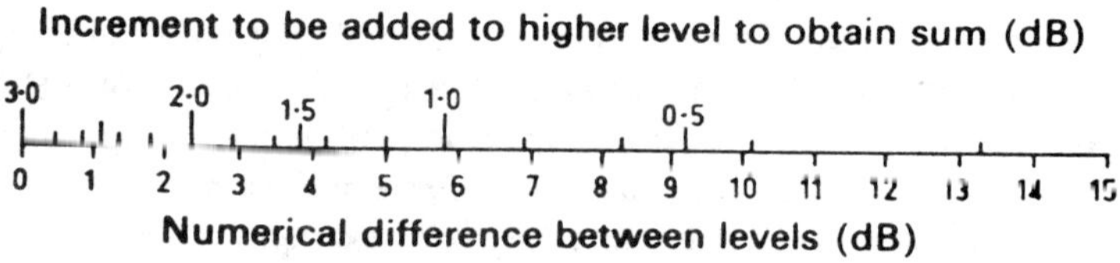

Figure 3-9 Graph for addition of two decibel levels.

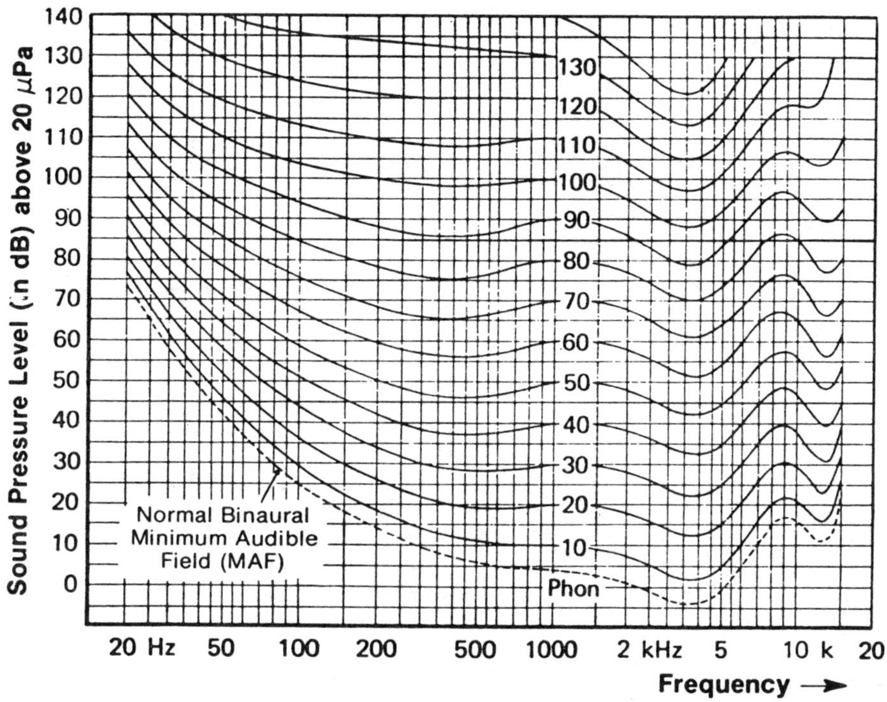

Figure 3-10 Normal equal loudness contours. They can be applied when:

a) The source of sound is directly ahead of the listener.

b) Listening with both ears.

c) The sound is measured in the absence of the listener.

d) The listeners are people with normal hearing aged between 18 and 25 years.

The ear also has a nonlinear amplitude response. An increase in Phon level of 10 dB represents an approximate doubling of perceived loudness. The frequency response of the human ear is the inverse of the equal loudness curves. Estimating loudness level on a simple instrument is done by making its frequency response have the same general shape as that of the ear's response. Sound level meters have standardized weighting filters with A, B, and C response curves as shown in Figure 3-11. These curves simulate the response of the human ear at low, medium, and high amplitudes. The original intention of the concept of the weightings was that the A scale should be used for measurement of sounds of low pressure level, while the B and then the C scale should be used at higher pressures. However, this was not completely satisfactory in practice, and in recent years the A scale has been most widely used as an indication of subjective loudness over a wide range of amplitudes and for evaluating hearing damage risk. Measurements in dBA are properly referred to as sound levels. Readings with a flat response amplifier are referred to dB SPL (sound-pressure level).

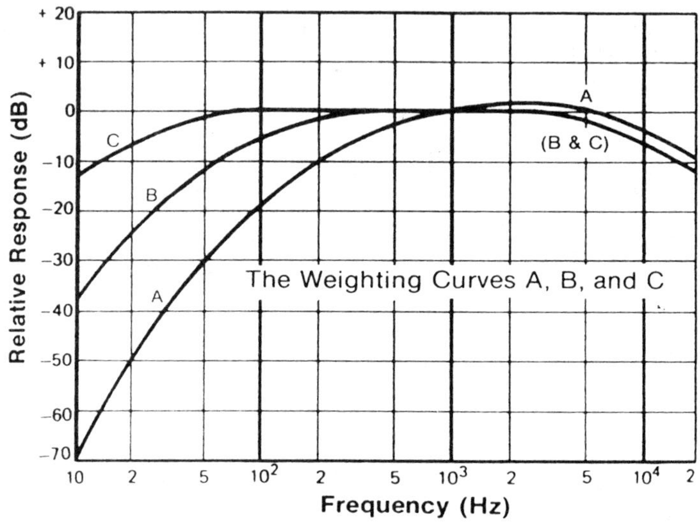

Figure 3-11 The frequency weighting networks.

COMPLEX NOISE

Sounds resulting from sinusoidal vibrations are pure tones, but the great majority of common sounds are of a more complex nature, having components at several or all frequencies in the audio range. For example, when sounds are produced from musical instruments, one fundamental frequency predominates, but others are present which are harmonically related to the fundamental frequency and which give character to the sound of the instrument. For convenience in the analysis of such sounds with many frequency components, the sound spectrum is commonly used. This is a graph of sound pressure amplitude versus frequency, averaged over a particular interval of time, in a pure tone which appears at a particular frequency as a line whose length is proportional to the amplitude of the sound pressure, as in Figure 3-12. Sounds with many harmonics, such as those of a musical instrument, or other periodic sounds appear as a number of lines at discrete frequencies (Figure 3-13). Spectra of this type are referred to as *line spectra*. The majority of common sounds, however, are nonperiodic and therefore have components at an infinite number of frequencies. These appear as a *continuous spectrum*, similar to the curve of Figure 3-14. Common examples of the continuous spectrum are the hiss of an air jet, or even the background noise in the room where we sit. Frequency analysis is more thoroughly discussed in a later chapter of this book.

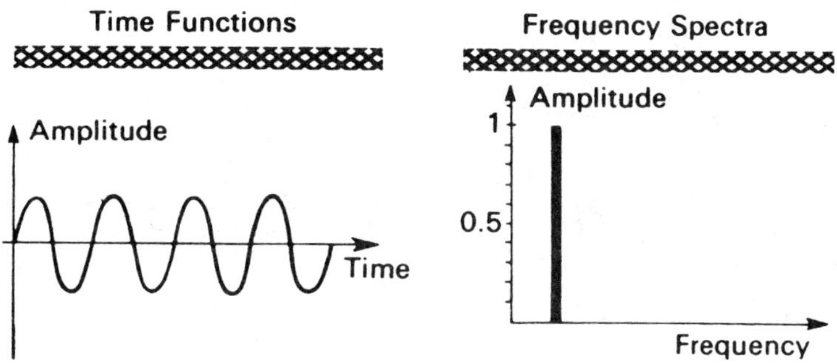

Figure 3-12 Sound Spectrum of a pre tone.

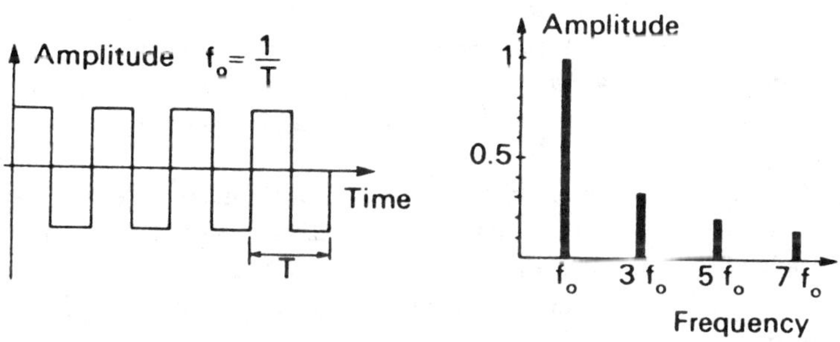

Figure 3-13 Sound spectrum of a periodic function.

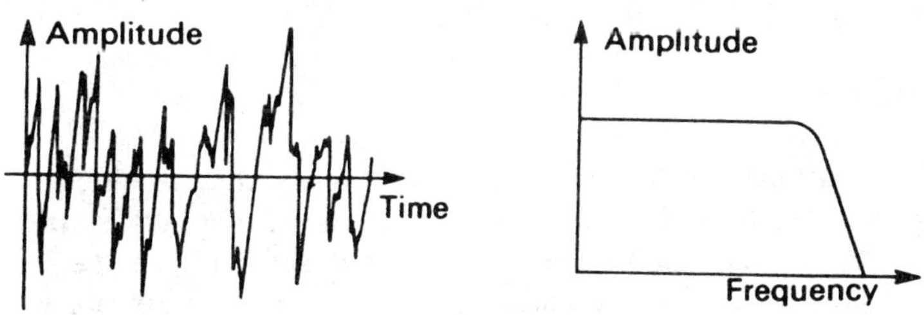

Figure 3-14 Continuous spectrum.

4 ENGINEERING CONTROLS AND DESIGN

Legal requirements make it imperative that steps be taken to reduce noise. When employees are subjected to sound which exceeds the permissible exposures set forth by regulations, feasible administrative or engineering controls must be utilized. *Feasible* means capable of being done, accomplished or carried out; capable of being dealt with successfully.

Engineering controls are those which reduce the sound intensity either at the source of the noise or in the hearing zone of the workers. The ideal way of controlling noise is at the source, to prevent noise-induced loss. To accomplish this, the source of the noise must be known as well as its transmission path.

There are many interacting elements which make analyses of sound sources and transmission paths in industrial plants complex problems. These elements include various shapes and sizes of enclosures and equipment, the presence of diversified vibrating, radiating, and reverberating surfaces, and differences in environments.

The following are engineering control measures which can be used to curtail noise: maintenance; substitution of machines; substitution of processes; vibration damping; infusion of noise level specifications when ordering new equipment; isolating noise sources; and isolating the operator. Table 4-1 lists typical industrial equipment noise sources.

MAINTENANCE

A method which may be obvious but many times overlooked is regular and thorough maintenance. This is essential in order to attain efficient noise control. A machine can be the best designed and built on the market, but vibration and wear may eventually cause it to become noisy. A good maintenance program should include adequate lubrication of moving parts, proper alignment, and proper balancing of all rotating members. All worn bearings, gears, clutches, and other components should be adjusted or replaced. This not only helps to cut down noise but also increases efficiency. Panel guards, mounts, and all other members which are somehow attached to a machine should be securely tightened to keep the vibration at a minimum. Fasteners should be checked periodically to make certain that they have not loosened. A lock washer helps prevent loosening and should be used wherever applicable.

Table 4-1 Industrial Equipment Noise Sources

System	Source
Heaters	Combustion at burners
	Inspiration of premix air at burners
	Draft fans
	Ducts
Motors	Cooling-air fan
	Cooling system
	Mechanical and electrical parts
Air-Fan Coolers	Fan
	Speed alternator
	Fan shroud
Centrifugal Compressors	Discharge piping and expansion joints
	Antisurge bypass system
	Intake piping and suction drum
	Air intake and air discharge
Screw Compressors (axial)	Intake and discharge piping
	Compressor and gear casings
Speed Changers	Gear meshing
Engines	Exhaust
	Air intake
	Cooling fan
Condensing Tubing	Expansion joint on steam discharge line
Atmospheric Vents, Exhaust and Intake	Discharge jet
	Upstream valves
	Compressors
Piping	Eductors
	Excess Velocities
	Valves
Pumps	Cavitation of fluid
	Loose joints
	Piping vibration
	Sizing
Fans	Turbulent air-flow interaction with the blades and exchanger surfaces
	Vortex shredding of the blades

Maintenance is an important part of engineering controls because it keeps machines running at maximum efficiency and noise at a minimum. Simple maintenance operations such as sharpening cutting tools can eliminate some noises as well as decrease cutting time. Cracks of any nature, whether in cutting tools or in machines, can produce noise and should be welded or the part replaced.

Faulty installation and/or maintenance can result in excessive vibration. Equipment should be periodically checked. Gradual increases in vibration should be examined in routine maintenance; sudden increases call for action. Increased vibration in machinery can be caused by the following:

- rotational imbalance which requires rebalancing
- misalignment of couplings or bearings
- eccentric journals
- defective or damaged gears
- bent shafts
- mechanical looseness
- faulty drive belts
- rubbing parts and resonant conditions

Rapid increases in vibration can be traced to a variety of causes, such as lack of lubrication, overload, or misalignment. In pumps, cavitation may erode an impeller. In fans, dirt may adhere to the blades and then break off unevenly.

MACHINES SUBSTITUTION

Machines are almost constantly running at their maximum speed. This sometimes produces sounds at very high frequencies which can be eliminated with larger, slower running equipment. The rate of production can be the same and even greater, if necessary, than that of the smaller machines. With larger machines, the noise is cut down; they also require less maintenance because they do not always operate at peak capacity. Presses can be substituted for hammers, immediately reducing the amount of exposure decibels. However, such substitution can be costly and may not always be economically feasible.

Considerably quieter hydraulic and/or pneumatic equipment can be used instead of mechanical equipment. Hydraulic power supply units can be placed on antivibration mounts. Also by cutting down on piping and number of fittings required and by using flexible hoses, noise can be reduced. Some manufacturers of hydraulic equipment are promoting use of an electric motor which operates at 1,200 rpm rather than 1,800 rpm. A larger pump operating at lower speeds is quieter and yet delivers the same flow as a smaller unit running at a greater speed. Hydraulic power units can also be placed outside a plant or in soundproof enclosures, thereby reducing noise to a bare minimum.

Machines, such as compressors, have had major noise problems with their exhausts. This type of noise can be reduced with mufflers or silencers. Many types of mufflers are used commercially and are discussed in a later chapter. The dissipative type uses sound-absorbing material; the reactive type has various geometrical shapes to provide impedance mismatch for the acoustic energy. Some mufflers combine both effects, and performance varies with the frequency of the sound and the backpressure created by the muffler.

Exhausts of air valves, cylinders, and other pneumatic devices have been equipped with silencers and are capable of reducing noise by 10 to 13 dB when operating under pressures of 80 to 100 psi. These units have hemispherical interference chambers made from sintered powdered bronze that permit contaminants and oil to pass through with air, while producing a minimum of backpressure.

Using belt drives instead of gears, noise reduction can be significant. The belt is one solid piece of material and makes relatively little noise. It runs between two pulleys under tension, whereas the gear drives using link chains are usually noisier. Gears have teeth in which the chain links must ride. The chain must be kept in tension and if tension slacks off, chain drives may jump a tooth and create a loud noise. The chain alone, just by rotating, creates more noise than the belt drive; therefore, the belt drive can be a substitute for gears.

PROCESS SUBSTITUTION

Substitution of a quieter process, machine, or tool is another method of controlling noise. Examples of this method are drilling for punching, pressing or rolling for forging, hot forming for cold forming, grinding of castings instead of chipping, and hydraulic presses instead of mechanical presses. Sometimes by substituting for a process, automation may be increased; increased automation reduces the number of employees to be exposed to noise.

Welding is often mentioned as a quiet substitute for riveting and is also faster than riveting for fastening two parts together. In riveting, sometimes the holes do not line up and they have to be redrilled. This is time consuming and induces added noise which could have been eliminated by welding. Welding is also longer lasting. Rivets, when exposed for long time periods, can corrode or rust, leading to loss of strength and eventual failure.

There are newer methods and machines available which are orbital, spinning, or radial cycloidal peening instead of impacting to form rivet heads or flares. By using these newer methods, riveting can be performed at reduced noises. A worker who is exposed to riveting all day is exposed to one of the highest noise levels on the dB scale. It is sometimes impossible to control noise at the source or path, and personal hearing protection is not acceptable as a long-term solution.

VIBRATION DAMPING

It is usually not practical or possible to isolate resonant vibrations from some metal structures such as thin panels, conveyors, and chutes. Instead, they are often given a damping treatment. Damping is a restraining force opposing or reducing the amplitude of a vibration into thermal energy. Damping, however, is of little value for other than resonant vibrations. Smaller panels vibrate less than larger ones; by providing air openings in large panels, radiated noise can be decreased.

Vibration isolator mats, pads, or sheets can be placed between machines and foundations or floors. These isolators can be single or multiple layers, depending on their use. These isolators are made of natural or synthetic rubbers, rubber-like plastics, cork, felt, sisal, metallic wool, dense glass fibers, and other materials. Pads made of a sheet of steel sandwiched between two layers of absorbent material within a sealed envelope of sheet lead are effective for high-tonnage presses and other impact machines (see Chapter 8). While pads are the simplest mountings, they may not be desirable because of restricted height adjustment, uneven contact with rough floors, or difficulties in protecting them from oil. Ribbed, waffled, or studded constructions are sometimes used for protection from oil, and bolts or wedges can be incorporated in the isolators for height adjustments.

Rubber isolators may be preferable to other types of isolators because of economy, the elimination of necessity for adjustment, the ease of installation and replacement, and the characteristics of the material. A rubber isolator's most valuable properties are its elasticity, which makes it a good shock absorber, and its ability to reduce noise transmission. Rubber has the characteristic of relatively large deformations due to loads.

Steel springs are also widely used, but generally for low-frequency isolation problems. With most common installations, springs are used to provide a natural frequency of the support mounting lower than that of the machine. In these cases the machine will pass through resonance with the support of startup or shutdown. For this reason, it is necessary to obtain a damping effect by using springs with other materials to limit movement as the machine passes through the resonance frequency. Adjustable snubbing can also be provided to limit movement, and is especially useful in limiting movement from unusual loading.

The success of noise control is based on reduction at the source, the transmission path, or protecting the receiver. Table 4-2 shows several common means of noise abatement. Each noise problem must be handled individually. There are no hard and fast rules. A problem may entail several malfunctions in a piece of equipment, and there may be more than one source. Probably the chief culprit in noise generation, and most important in transmission, is vibration. This can be the most detrimental of noise generators since it can cause machines to deteriorate faster, produce higher maintenance costs, and lead to production rejects.

EXCESSIVE VIBRATION CAUSES

Imbalance is the most common cause of vibration. Loose mountings or holding devices may be a source of trouble. Differences in dimensions of mating parts, such as a shaft with a diameter that may only be minutely smaller than the corresponding bore of the rotor to be mounted, is another possible cause. Others include bent shafts, which locate the off-center from the axis of rotation, and machine errors where less material is removed from one side of a roll than from another.

Table 4-2 Control Methods for Handling Excessive Noise

Plenums	Admit low-velocity cooling and combustion air to noise sources. At the same time they prevent the escape of excessive noise.
Enclosures	Consist of a hard outer shell and adsorbent liner for trapping noise.
Lagging	Lining that absorbs radiated noise. It is directly applied to piping, vessels, valves, and other equipment.
Damping	Reduces the amount of noise generated by vibrating surfaces.
Silencers or mutes	Attenuate noise from high-velocity flows of gases. This is done by multiple reflection of sound waves from acoustically absorbent surfaces, elimination of turbulent flow, reduction of flow velocity, and cancellation of sound waves by flow interaction without employing absorption.
Ear plugs and muffs	Serve only to protect workers in the immediate vicinity of the source. They should not be considered a permanent protection, only as an expedient.

Assembly errors constitute another common cause of vibration. They can be arrested by insuring surface-to-surface contact of mating parts. On fans and blowers, rebalancing should be done after worn-out fan blades have been replaced.

Misalignment can cause excessive vibration. Flexible couplings should be aligned as carefully as solid couplings. Although misaligned flexible couplings transmit torque, vibration can be so high as to cause excessive wear of bearings and journals. Proper lubrication of flexible couplings will usually correct this.

Drive belts are another source of vibration. The length and tension on a belt can generate resonance within the drive belt, causing it to behave like a vibrating string. A high axial vibration and excessive thrust bearing wear are caused by unequal tensions in multiple V-belts. The problem can be handled by replacing the belts with ones of equal tension. Proper alignment of belts will also help.

Gear problems can stem from unbalance, resonance, tooth wearing, and bearing instability. Usually, lubrication has no effect on the noise generation of a gear system. The type of oil and its viscosity are of small importance as long as contacting surfaces are coated and a dry running situation does not exist. The use of lightweight viscoelastic materials may help, but this is primarily a problem not easily handled.

Antifriction bearings contribute to noise by the out-of-round shape of the bore or shaft involving the bearing-pressed fit. This may be alleviated by utilizing a slip fit between mating surfaces. Adhesive is used to keep the bearing race from turning in the matching bore. Figure 4-1 shows noise level ranges for various operations.

ISOLATION AND DAMPING OF VIBRATION

Reduction of vibrational forces can be handled through isolation (a separate maintenance procedure), general maintenance and modification of existing systems, and damping. General maintenance and modification procedures have been previously discussed. Maintenance modification methods can reduce vibration and sound transmission by

- replacement of worn parts
- proper realignment and balancing of the rotating equipment
- proper and periodic lubrication to reduce frictional forces
- tightening of parts
- proper assembly of equipment
- reduction of equipment operating speed
- reduction of flow velocities of gases and liquids which will decrease turbulence and speed
- periodic cleaning of equipment

Isolation should be considered a key step for noise abatement and machinery operation performance. The isolation of vibration in some cases offers virtually a complete solution to the problem of noise transmission. This is particularly evident in cases of transmitted noise to the floors below the machinery or to rooms adjoining a mechanical operation. If the system is excited by a continuous vibration at one of its natural frequencies, a condition of resonance results. Under such conditions, large amplitudes of vibration are induced by very small amounts of energy. The vibration transmitted to a floor on which a particular machine is mounted may set up sympathetic movement of other machinery installed elsewhere on the floor. Decoupling of machinery from the floor with the use of vibration isolators will alleviate many problems associated with vibration and noise transmission through structures. The use of vibration isolators will also alleviate structure-borne noise problems associated with vibration. Proper selection will ensure the desired degree of isolation efficiency. Manufacturers of vibration isolators can assist in selecting proper isolators which range from cork, the original vibration and noise isolation material, to more sophisticated air-mount systems which have lower natural frequencies and thus higher isolation frequencies.

Proper application of damping materials such as plates, shells, and webs will reduce resonances that are generated in their structures. In selecting the proper damping materials, one must examine the type of structure to be damped,

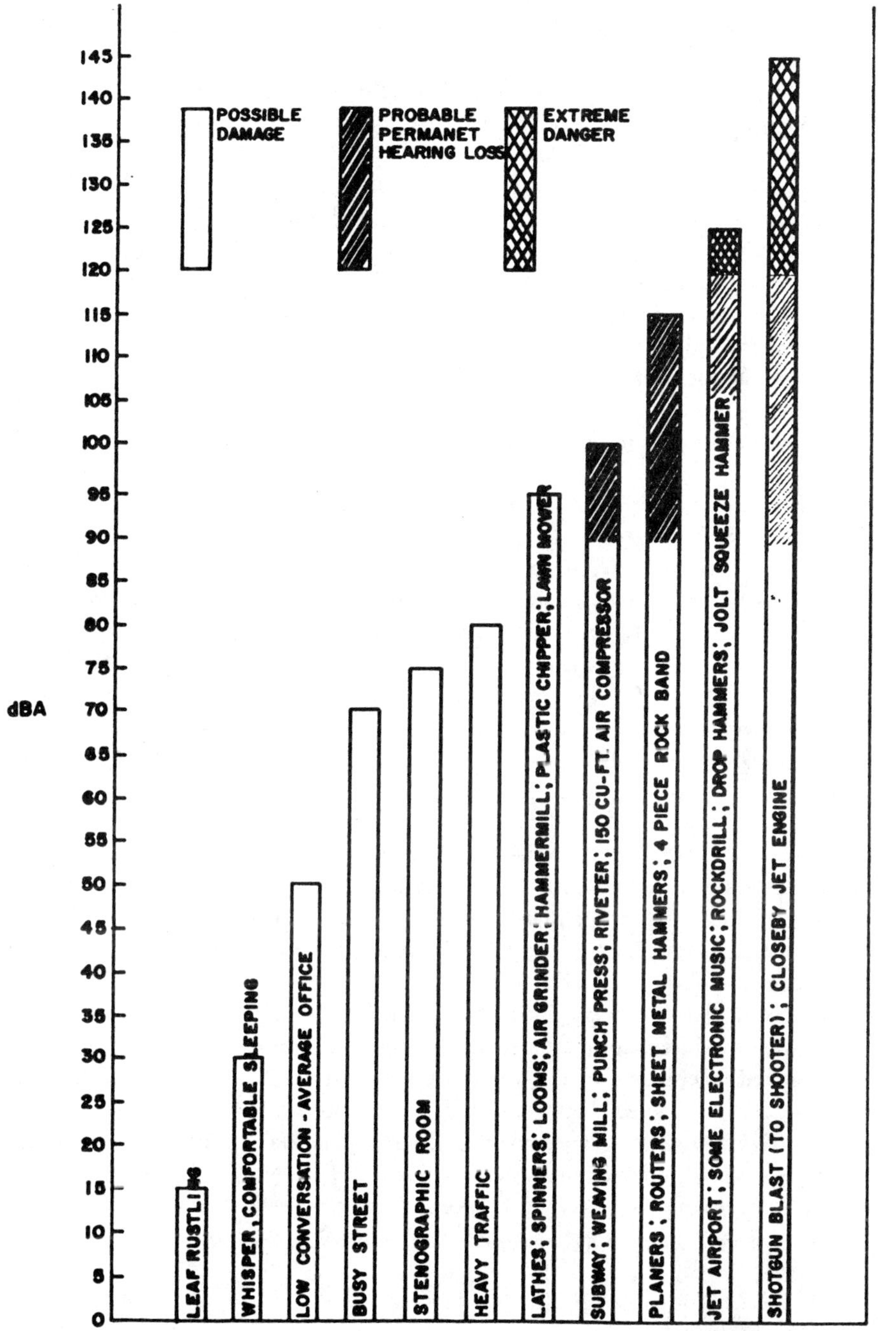

Figure 4-1 Noise-level ranges for various operations.

the frequency of vibration, operating temperature, function of the vibrating part, size of equipment, the sound level generated, and the size of the area exposed to noise and vibration.

Some of the more common materials include lead, sand, impregnated felt, and loaded asphalts. These can be obtained in sheet or mosaic form. Good damping materials should display:

1. Stiffness that is compatible with the stiffness of the structure being damped. The purpose of this is that the maximum amount of energy is extracted during bending motion.
2. Temperature resistance. Careful consideration as to operating temperatures must be given.
3. A thickness at least equal to the section of the material being damped. Sometimes several times the thickness is needed in order to obtain maximum damping.

INCLUSION OF NOISE-LEVEL SPECIFICATIONS WHEN ORDERING NEW EQUIPMENT

Quieter machines are generally more expensive—on the average, their estimated cost is about 10 percent higher. However, it is almost always more economical to buy quieter machines than to try to reduce noise by modifying the machinery after purchase. Purchase orders should specify maximum permissible noise levels for the equipment. Major problems being encountered concern economical procurement of equipment that generates less sound and still performs the desired functions efficiently.

OSHA guidelines discussed elsewhere in this book make it almost mandatory to include noise level specifications when ordering new equipment. In the long run, money is actually saved by buying the quieter machines. Everyone profits from the quieter machines. The employees' hearing is better protected, work is performed more efficiently thus increasing production, and the employer gains from the increased production.

Since the sound produced by a machine generally increases with its horsepower, the lowest power needed for a specific application should be used. Higher-speed machines are also usually noisier, so the lowest speed required is recommended. However, improving dynamic balance can reduce noise levels at higher speeds. Operating speeds should be kept as far as possible from the resonant speeds of the machine. Acceleration and deceleration forces, as well as the mass of moving parts, should be kept to a minimum. Machines which are built with noise control specifications usually have tighter tolerances.

ISOLATING NOISE SOURCES

Controlling noise at the source is the ideal way of preventing noise-induced hearing loss. The results are long lasting, the individual operator and co-workers are protected, and the need for protective equipment is eliminated. A method of controlling noise at the source is to build an enclosure around the noise source. Sources can range from small pumps to very large machines, and in considering an enclosure the key to control is proper design. Factors such as environment, temperatures, durability, accessibility, personnel contact, appearance, weight, and portability are all important. If an enclosure is built around an entire machine, the machine must have easy access in case of emergency. Also, if the machine must be periodically lubricated, accessibility is also important. An enclosure around a machine can be built with soundproof panels or soundproof doors. Also provided in the enclosure can be glass panels or doors through which vital moving parts of a machine can be observed as well as gauges and other instrumentation.

Enclosures for equipment must be constructed of proper sound-damping materials. Lead is probably one of the oldest and most common sound-attenuation materials in use today, and is still popular for many applications. In general it has two basic areas of application in sound control as a barrier used to block sound paths and as a damping material for structures and surfaces to reduce transmission of vibration. Within each of these areas, there are many variations and specialized uses of lead and mixtures of lead and other materials. Perhaps the primary reasons for its continuing popularity are its extreme versatility and simplicity—even for the uninitiated. It can be cut with a common scissor, molded and shaped to any contour, and is unaffected by solvents, coolants, cutting oils, and other materials commonly used in industry. It can readily be laminated to other materials such as wood, plastics, and steel, and is often mixed with other materials such as epoxy and plastic to produce an effective sound absorbent with a broad range of properties. For example, lead powder mixed with various plastics and epoxies forms a compound that can be troweled on resonating machinery guards and shrouds to form a thick sound absorbent that retains flexibility and resiliency. It can be sprayed, bonded, or coated onto or into metal parts and machinery to reduce and contain sound at its source.

Lead powder in vinyl with a fabric backing is commonly used as a curtain or shroud to enclose noisy operations and processes. Used as an isolation pad under equipment and even structural components in buildings, lead is unsurpassed as a noise insulator. However, lead as a sound control is perhaps most widely used in building construction applications.

A glass barrier also offers the potential for controlling noise while providing visual communication. For the same thickness, it is a better sound barrier than most brick, tile, or plaster. An average noise insulation of 35 dB can be obtained by using a single-leaf wall weighing about 6 pounds per square foot. However, single-leaf wall construction is not efficient if an average insulation of 40 dB or more is required. In this case a double-leaf type of wall construction must be used.

Isolating noise sources is one of the best methods for controlling noise. This technique should be used wherever possible and feasible. But there are many instances where the noise source cannot be isolated. In this case, noise should be controlled in its transmission path or, as a final step, the operator can be isolated.

OPERATOR ISOLATION

As a solution to noise control in some cases, the operator must be isolated. There are some areas in which noise levels are much too high and cause severe hearing loss in a short period of time. The operator might then be isolated in a soundproof booth. The booth can be made as large or as small as necessary. If the operation is simple, a small booth will do the job. All the controls required by the operator can be placed inside the booth.

When the noise level within a confined area is too high to allow workers into the area even with the use of personal protection devices, the operation might be automated. An automated process can be supervised from observation posts, that is, from remote-control stations where the workers are adequately protected. A remote-control post can receive information via closed-circuit television, or it can be a highly insulated area within the department. The production procedures are handled by mechanical devices under operator or computer control. Rolling mills are controlled from completely soundproof cabins. The same is true of workshops used for assembling and testing engines. The noise of such operations normally cannot be stifled at the source.

Engineering controls are effective in noise control and should be used wherever applicable. The correct engineering control should be used to obtain efficient results. By reducing noises, the operation is usually increased and machines run much more efficiently. Workers actually work at greater productivity with less noise.

OSHA is forcing employers to use engineering controls. New plants are being designed with noise control in mind which will result in savings as well as implement safety in the long run. Machines are also being designed to run more quietly. Engineering control and future engineering technology will help to keep noise at a minimum.

SYSTEMS DESIGN

In designing a system for reducing noise, engineering controls should be used wherever applicable. These controls can be effective if used properly. Systems, however, are much more complicated and must be given much more consideration. Instead of just using engineering controls to quiet noise, the placing of the machines is also a vital factor. If machines are layed out too closely together, the operator may be exposed to an excessively high dB level, whereas if spaced adequately apart, noise levels will be within acceptable limits.

In designing a system for office areas, careful attention should be given to the layout design, creating an environment that will attract and keep employees. Sound control for ceilings in offices must be planned at the architectural stage. It becomes very expensive and time consuming to soundproof an office after it is built.

Wall and ceiling construction should provide approximately the same degree of sound control through each assembly. Where a drywall partition is used for sound isolation, the construction should extend from slab to slab. Sound can travel through a suspended acoustical ceiling, up and over a partition attached to a suspended ceiling, and into other areas.

Optimum sound isolation requires that the integrity of drywall partitions and ceilings not be violated by cutting holes for vents or grilles or by recessing cabinets, light fixtures, and so on. Instead, they should be surface mounted. Door and borrowed light openings are not recommended in party walls. The septum layer in the triple solid drywall partition should always remain whole.

Where holes are necessary, avoid placing them back to back and immediately next to each other. Electrical boxes should be staggered, preferably at least one stud space. A nonhardening, nonskinning resilient caulking material should be used to seal all cutouts, such as around electrical and telephone outlets, plumbing escutcheons, and wall cabinets. The backs and sides of electrical boxes should also be caulked to prevent sound leaks. Caulking should be used to seal all intersections with the adjoining structure, such as under-floor and ceiling runner tracks, around the perimeter at which the assembly meets the floor, ceiling, partition, or exterior wall at a vertical intersection occurring at columns and window mullions.

Avoid construction such as ducts, rigid conduits, or corridors, which act as speaking tubes to transmit sound from one area to another. Common supply and return ducts should have sound attenuation lines. Conduits should be sealed. Doors leading to a common hall should be gasketed around the perimeter and should not contain return-air grilles.

To isolate structure-borne vibrations and sound, resilient ceiling systems and floor coverings are recommended. Vibrating or noisy equipment should have resilient mountings to minimize sound transfer to structural materials. Ducts pipes, and conduits should be broken with resilient, nonrigid boots or flexible couplings where they leave vibrating equipment; they should be isolated from the structure with resilient gasketing and caulking where they pass through walls, floors, or other building surfaces.

Since production-area ceilings are high, office walls are often constructed of bricks or acoustical partitions. The ceiling is then suspended from the underside of the plant roof. The office walls will usually prove to be good sound barriers. However, the sound-deadening capacity of the ceiling will be far below what is required for conversation, even when acoustical ceiling tile is used in the T-bar suspension systems. The ceiling will control sound originating with the office, but will not serve as a barrier to external noise.

Noise from production equipment can reflect off the underside of the plant roof and focus on the acoustical tile ceiling of the office. As a result of this unique focusing problem, noise levels within the office can be more distracting than in the plant.

Constructing full-height insulated office walls to the underside of the plant roof is usually uneconomical in high-ceiling areas. The high cost of materials and labor, interruption to production, and modifications to fire-control sprinkler systems usually make high-wall office systems prohibitive in cost.

Using a lead powder-loaded, vinyl-coated fabric can help reduce outside noise substantially. The material is limp, dense, and nonporous and is easily cut with a knife or scissors. It weighs about 0.9 pounds per square foot and is approximately 1/16-inch thick. Most important, the material can be installed easily over one office in an entire group, and its effectiveness evaluated before making a commitment for a total installation.

The lead-loaded fabric can be installed by plant personnel, quickly and easily, usually with only oral instructions. Rolls of vinyl are unrolled on the floor, cut to length, and laid directly over the T-bar suspension of the ceiling. No additional support is necessary. Noise levels ranging from 95 to 100 dB in offices have been reduced after installation of lead-loaded fabrics to 75 to 80 dB.

5 PERSONAL SAFETY

Often it is not possible to stop the generation of noise at its source or by the use of acoustical materials and noise barriers. In such cases, individual protection against the troublesome noise is the only alternative.

ACOUSTIC PROBLEM

Individual protection against noise is accomplished by the use of ear protective devices. The function of such devices is to place a barrier between the noise source and the hearing mechanism within the ear. This barrier is a physical device which is either worn over or inserted into the external ear. This barrier will reduce the amount of noise energy transmitted to the inner ear. The degree of protection available through such devices depends on a number of physical and physiological factors that control the behavior of the sound energy as it meets the protective barrier. Figure 5-1 illustrates the acoustic problem by a block diagram. Figure 5-1, however, does not fully illustrate the complexity of the factors which determine the attenuation characteristics of the barrier because sound energy reaches the inner ear by three different paths. These three paths are through vibration of the barrier, by acoustic leaks through or around the barrier, and through bone conduction which in effect short circuits the barrier. The importance of these pathways depends on the frequency and intensity of the noise and the amount of leakage around or through the protective device. Figure 5-2 illustrates a block diagram of these three paths.

Where the noise airborne path is blocked completely, the noise attenuation afforded the inner ear is limited to approximately 50 dB below the environmental level. This attenuation limit is the result of the conduction of surrounding noises by the skull to the inner ear. One can see how bone conduction varies with frequency in curve C of Figure 5-3. In general, a 50-dB attenuation is more than satisfactory for sufficient ear protection in an industrial environment. Therefore, the limitation due to bone conduction is neglected, except in very extreme cases.

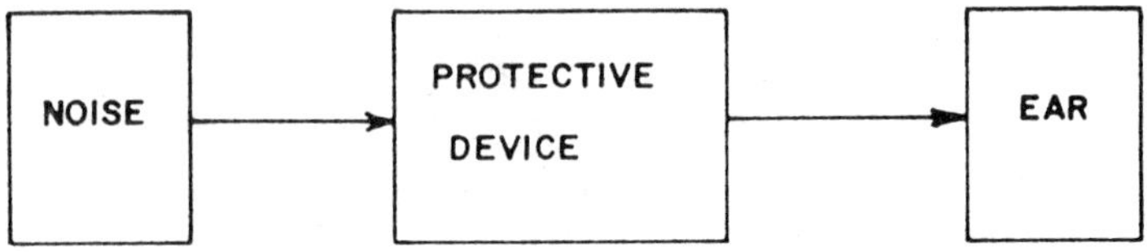

Figure 5-1 The acoustic problem in general.

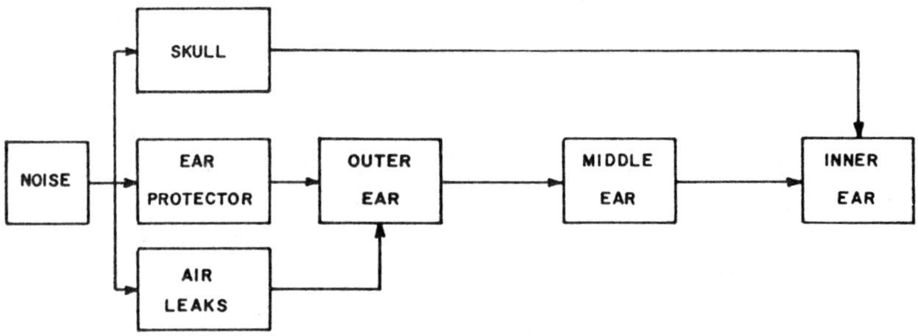

Figure 5-2 The three different paths by which sound energy can reach the inner ear while a protective device is being used.

Leakage, on the other hand, cannot be neglected and must be corrected. Curves A and B of Figure 5-3 illustrates the attenuation losses due to leakage into an improperly sealed ear muff. A poor acoustical seal greatly affects the attenuation characteristics of an insert-type ear protector (see Figure 5-4).

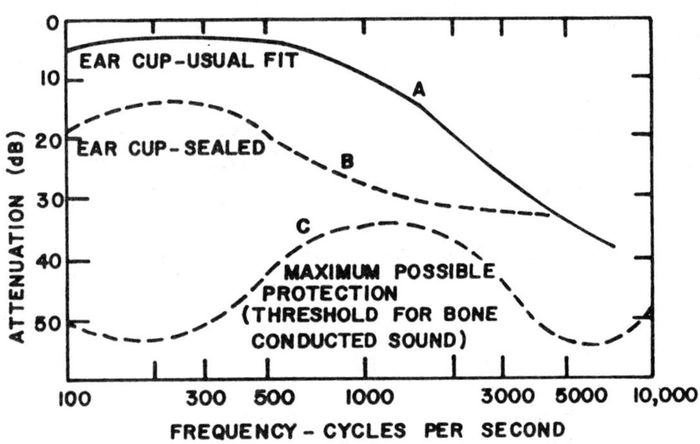

Figure 5-3 Variation of bone conduction with frequency.

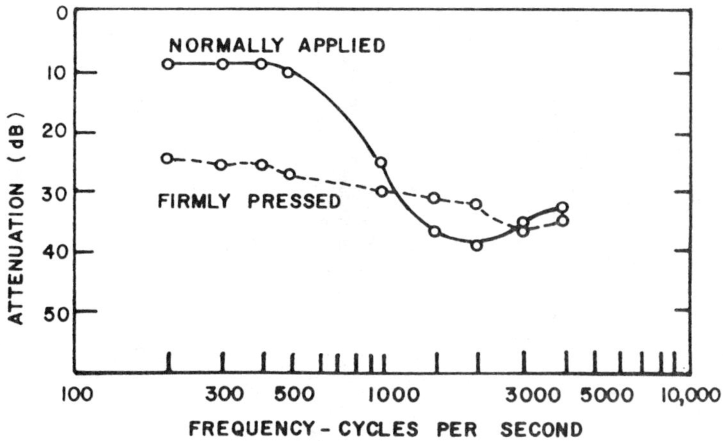

Figure 5-4 Effects of poor acoustical seal on the attentuation quality of an insert-type ear protector.

EAR PROTECTION

Surveys of industrial environments reveal that dangerous noises are generally classified in two separate categories: the moderate-level to high-level noises of most common manufacturing areas, and the extremely noisy areas such as jet engine testing areas. A large percentage of the overall noise levels in the general manufacturing industries averages around 100 dB. In the extremely noisy industries, levels reach in excess of 130 dB.

The noises which fall into the two general categories have their own individual characteristic frequency spectra. For a particular exposure to any noise, the characteristic noise spectrum should be compared with the adopted damage risk criteria to determine if a protective device will be expected to give adequate ear protection. Figure 5-5 illustrates the noise levels which can be tolerated by a partially protected ear.

By adding the attenuation expected of the protective device to the criteria, the highest noise levels at which typical protective devices afford sufficient protection can be determined. Therefore, Figure 5-5 indicates that a relatively good ear protector will be sufficient for most of the noises in the general manufacturing industries. However, in extremely noisy industries or operations, a combination of several protective devices should be used to ensure adequate attenuation. In addition to providing adequate attenuation, an ear protector must be nonirritating to the skin, comfortable and sanitary over long periods of time. There are two general types of ear protectors used widely by industry: the cup or muff type and the plug or insert type. Figure 5-6 illustrates some of these protectors.

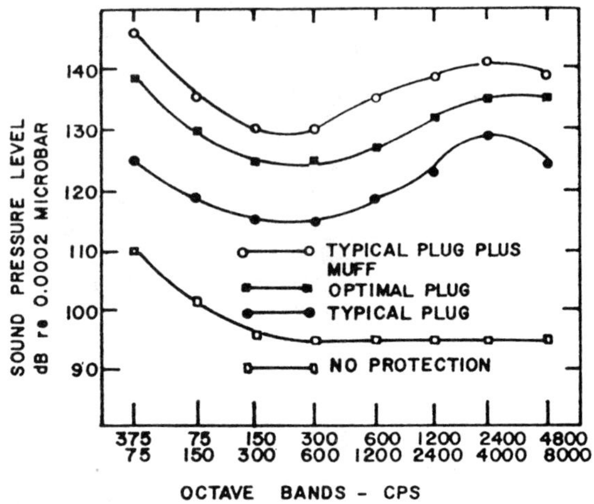

Figure 5-5 Tolerance levels of the human ear while protected in different ways.

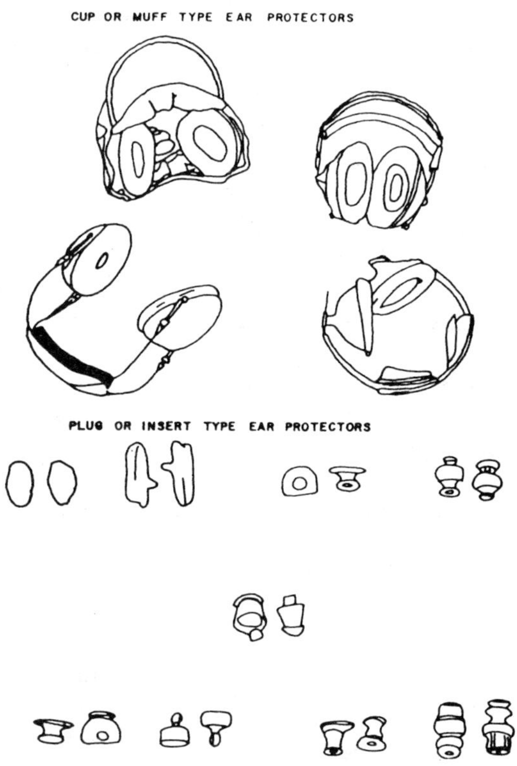

Figure 5-6 Various types of ear protectors.

Insert Type

Ear plugs are inner-ear protection devices which are designed to occlude the ear canal. They are made from soft materials, particularly rubber, neoprene, wax, cotton, fiberglass, or plastic. The choice between plugs and muffs depends greatly on the work situation and the employee's preference. Both types offer adequate protection at frequencies above 1,000 Hz. If an employee uses a helmet, he or she would probably be more comfortable with ear plugs.

There are several problems encountered by workers using ear plugs. In order to get a good acoustical seal against the sensitive inner lining of the ear canals, the inserts have to apply a certain degree of pressure. This pressure can cause discomfort and is the main reason why workers find them objectionable. Many employees often prefer to use cotton as a sound suppressor; however, this is a poor alternative. Figure 5-7 illustrates the attenuation level of pure cotton. Most ear plugs provide more than double the protection of cotton. Employers should discourage employees from improvising any type of inserts.

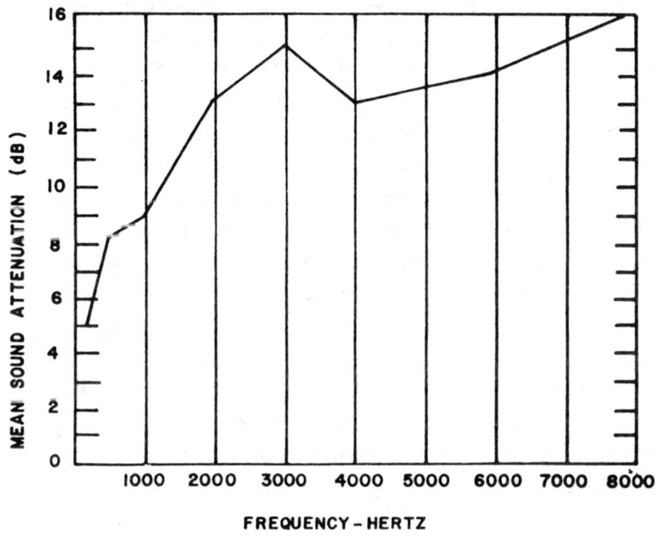

Figure 5-7 Attenuation levels of pure cotton.

Another problem is that ear plugs usually pick up a varying amount of dirt. Wearing ear plugs for an extended period of time may cause a "plugged" feeling, dizziness, or vertigo. When such a situation arises, the worker usually places the plugs on the work area where they can pick up dirt, metal filings, germs, and other debris. The only way to combat this problem is to fully educate the employee. It must be shown that it is necessary to clean the plugs every time they are to be used, unless taken directly from their container.

Because of the simplicity of design, ear plugs can be readily cleaned of any dirt they pick up.

Another problem is that ear plugs might lose their efficiency during the day, due to a break in the acoustic seal between the ear and insert. This occurs because of jaw movements which tend to change the shape of the ear canal. This problem has been remedied to a degree by utilizing a headband connecting the ear plugs. Lightweight headbands provide a small pressure on the canal caps to hold the plugs in place. When the band is properly adjusted to fit the head size and the plugs are properly inserted, the caps require less reseating after short periods of use.

Ear canals of different people vary in size. Most manufacturers provide standardized inserts in different sizes. However, this may cause some problems in fitting. The best method of effective control is to obtain personally molded ear plugs. There are kits sold on the market with soft putty-like material that can be molded in the individual's ear canal. This method thereby eliminates excessive pressure against the sensitive canal walls and gives maximum comfort. There are several problems, however, which can arise even when using molded types, one of which is that air can be trapped in the ear when first molding the inserts. This usually causes voids in the plugs which will prevent a good acoustical seal. It is very important that the plugs fit properly. Even the slightest leakage will lower the amount of attenuation as much as 15 dB at some frequencies. Some materials also might shrink when they harden, therefore eliminating the seal. One should evaluate all literature on such devices. Custom-made inserts should be fitted by trained technicians, nurses, or doctors.

The use of personal ear protectors involves more than merely purchasing and fitting. Employees' ears should be examined and tested before fitting with ear protectors; plugs should be designed and fitted individually for each ear. Workers should be educated to the necessity of using these devices and their care. There may be a certain amount of discomfort on the part of the wearers when they first use the ear protectors, and they should be told of this prior to issue. To obtain better acceptance, employees should be given a choice of several different styles of ear protectors. One of the most important factors is to purchase the type which will be worn most effectively.

MUFF-TYPE PROTECTORS

Ear muff protectors give over-the-ear protection against dangerous high-frequency noises. However, they should also allow the wearer to hear low-frequency sounds such as voices and warning signals. This poses a problem in hearing protection equipment. Workers have to be able to communicate with each other normally, or they may not use the devices. In choosing the proper ear muffs, an engineer must determine the sound levels generated in the work area. Such data may be obtained by using sound-measuring meters. Then the attenuation levels of the chosen ear muffs must be carefully determined.

Attenuation is a measure, in decibels, of how effective the device is in diminishing harmful high-frequency noises. The engineer can compare hearing protection devices by comparing the attenuation data that are usually supplied by the manufacturers. Figure 5-8 illustrates a common attenuation graph. An ideal muff should be highly effective in the high-frequency ranges because it is at these frequencies that most hearing damage occurs. Attenuation provided by ear muffs varies due to differences in size, shape, seal material, shell mass, and method of suspension. Another important aspect of attenuation is the wearer's head shape and size. The cushion between the shell and head also has a great deal to do with the attenuation frequency.

Many types of ear muffs and cups are available. There are basically four parts to an ear-muff device. The ear cups are the outermost covering designed to deflect as much noise as possible. These should be made of a rugged material capable of withstanding any mechanical shock. Although many manufacturers indicate that ear cups are shaped to fit all ear sizes comfortably, this is not always the case. Each worker should be fitted individually. Another part is the sound-absorbent ear-cup insert. The outer ear cups are lined with a

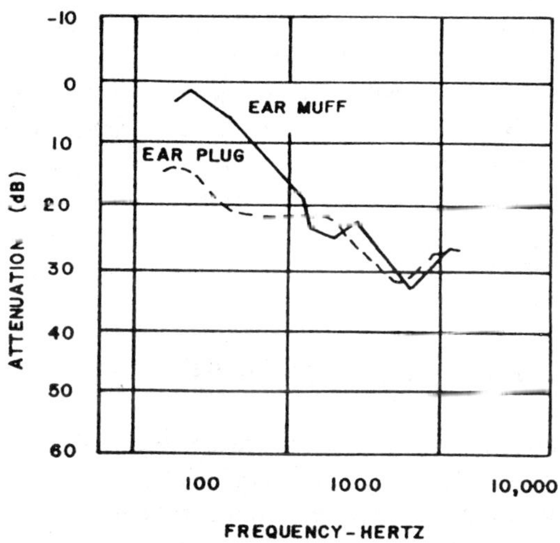

Figure 5-8 A common attentuation graph.

soft foam, such as a vinyl or polyurethane foam. This lining insert serves two purposes: (1) it provides additional noise reduction, and (2) it eliminates the "sea shell" effect of roaring noise generated when one holds an unlined cup over the ear. A third part is the sound-tight ear cushion, which is optional on some ear muffs. This reduces sound levels and prevents sound leaks by acting similarly to a washer and is usually constructed of a tough vinyl cushion. The fourth part is the headframe which connects the two ear muffs and suspends them from the head. The frame is easily adjusted with a type of slide-lock mechanism. It should mechanism. It should be light and have adequate tensile strength. One of the most important functions of a headframe is that it will provide a comfortable fit. It is necessary to provide a headframe that offers swivel action around the cups to eliminate bulkiness; in this way the wearer can adjust the headframe to provide the most comfortable fit.

Ear muffs also pose another problem; namely, they may interfere with other head gear. The swivel-action yoke headframe permits the ear cups to be worn with hard hats, bump caps, and welding helmets while other models can be fastened to safety caps. A spring tension band holds the complete unit on the cap with minimal pressure; the ear cups grasp the underside of the cap brim.

The most important aspect of personal protective devices is that they depend on an effective seal between the skin and the ear protector's surface. If even a small sound leak is present, the effectiveness and purpose of the ear muff is destroyed. A common disturbance encountered by users of ear muffs is that the muff might become loose as a result of talking, chewing, or simple movement. There is nothing that can be done to combat this problem except to reseat the cups occasionally.

Sanitation is another aspect in using ear muffs, although this problem is more typical of inner ear protectors. It is most important that the unit be easily disassembled for cleaning and sanitizing.

Most commercial ear muffs offered on the market give about the same degree of protection. The best device is one that is accepted by the worker and is worn properly. Personal fittings should be made to ensure comfort and effectiveness. One way of determining whether ear cups are performing effectively is to conduct hearing tests on workers before and after use. When properly worn and maintained, ear muffs can provide adequate protection against most industrial noise exposure.

COMBINED USE OF MUFFS AND INSERTS

If maximum protection for severe noise is required, both muff-type and insert-type protectors can be worn at the same time. The attenuation given by both will not be the sum of the two devices individually; however, a little less than this value can be achieved. The difference between the upper and lower curves in Figure 5-5 indicates the noise reduction that should be expected. In no case will the total attenuation be greater than approximately 50 dB, because at this point bone conduction becomes significant.

COMMUNICATION WHILE WEARING EAR PROTECTORS

Communication is essential in any industrial environment; therefore, an important aspect of an ear protector is its ability to allow maximum communication in conjunction with adequate articulation protection. In Figure 5-9 we see the degree of speech intelligibility articulation that is expected from

normal

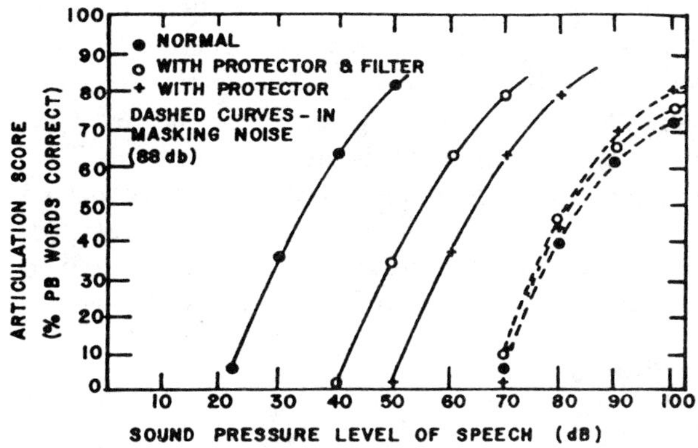

Figure 5-9 Degree of speech intelligibility with and without ear protectors.

normal ears—both with and without ear protectors—in several background noise levels. The devices used in this investigation were soft plastic insert types which have high attenuation values throughout the audible spectrum. The curves indicate that in steady-state background noise levels, up to about 85 dB, speech intelligibility is the highest when there is no protection. In background noises above 85 dB, speech intelligibility is gained if ear protection is worn.

Figure 5-10 illustrates the results of a study in which the intelligibility scores obtained by a subject wearing protector, were compared to the scores obtained

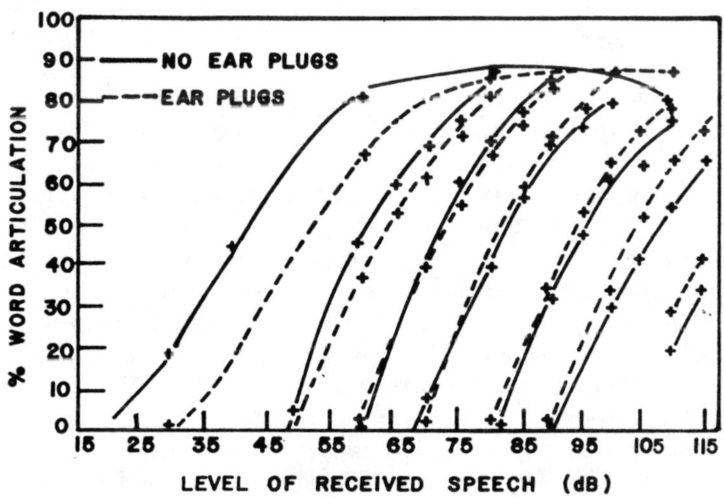

Figure 5-10 Results of a study of intelligibility scores.

obtained with protectors having speech filters. The protectors that had speech filters were constructed to give minimum attenuation of low frequencies where speech components are present. Speech intelligibility was tested in quiet surroundings and then in noisy surroundings. The results of these studies show that there is little difference in speech intelligibility scores with or without ear protectors with background noises of approximately 88 dB. In background noises below the level of 88 dB, intelligibility is highest without protectors; but in noises above 88 dB, intelligibility is better with the protectors. The protector with the speech filter indicates better articulation scores than the broadband attenuator only in background noises below the 88 dB level.

EAR PROTECTION PROGRAM

To institute an effective ear protection program, the industrial hygienist must educate the management and the employees to the complete noise problem before proceeding to define the benefits of ear protectors. Management must recognize that excessive noise produces hearing loss and, if high noise levels cannot be reduced, ear protection represents an effective solution to the problem.

The worker is entitled to an explanation of the noise problem in the working environment. It is important that this explanation does not produce apathy. Employees should understand that it is unlikely that total hearing loss will result from excessive noise exposures, but partial hearing losses are quite possible, and in the early stages these losses are not readily detectable. Emphasis should be placed on the fact that a good ear protector will give sufficient protection for all but the most extreme noise exposures. The use of makeshift ear plugs such as cotton should be discouraged because they do not always provide sufficient protection. It should also be emphasized to the employees that in the typical factory environment the use of ear protectors will not deprive them of necessary hearing capacity.

It is preferred that the distribution of ear protectors lies within the responsibility of the medical department. An examination of the ears should be carried out before the fitting of ear plugs. Any disapproval of their use should be noted and recommendations made to supply the employee with muffs or the assignment to a quieter working area.

So that the employees will get the maximum benefit from an insert-type protector, it is very important that they be fitted with the correct size. After being fitted, the employees should be given instructions on sanitation. Permanent ear protectors should be cleaned daily with soap and water so that all dirt and wax deposits are cleaned off before the plugs are reinserted. Workers should periodically examine their ear protectors for any defects. If any devices are found to be unsatisfactory, they should immediately be replaced.

The medical or safety department should have available for distribution two or more satisfactory types of ear protective devices. This has a psychological advantage in that it allows the employees to choose for themselves the device which gives them the most comfort and protection. The measurement of ear protector efficiency is a complicated and expensive procedure that industry cannot economically perform on every employee. The most practical method of determining the effectiveness of an ear protector is by periodic audiometric examination of the workers who use protectors.

During the first year of a hearing conservation program, a number of audiometric examinations should be conducted. Afterwards, hearing tests should be conducted along with periodic physical examinations. If an audiometric examination indicates a significant threshold shift, an investigation into the cause is indicated. The investigator should determine whether (1) the employee is regularly using the ear protector; (2) the noise exposure has changed; (3) the ear protectors are fitted properly; and (4) the individual is noise sensitive.

Supervision is a key factor to the success of an ear protection program. While the general guidance may be the task of the medical, safety or industrial hygiene personnel, supervision should be the responsibility of the plant supervisor. Supervision is particularly essential in locations where the insert-type protectors are employed. Here, the supervisor is the only person who is continually close enough to the employees to see that the protective devices are being used. The supervisor is one of the most important individuals in an ear protection program.

6 ENCLOSURES AND BARRIERS

A practical and efficient method of reducing noise in a system is to enclose it, thus cutting off some of the sound waves. Enclosures can be constructed from most common building materials. However, the degree of noise reduction depends on the surface weight, as well as the internal damping and stiffness of the material.

LEAD AS A NOISE BARRIER

A wide selection of products and materials is available for noise reduction, one of which is lead. Lead is used to combat noise pollution in a variety of applications. The metal is applied alone and in combination with various other materials to meet OSHA regulations in numerous industrial, business, educational, and residential complexes. Table 6-1 lists types of lead-containing materials that are used for industrial noise control. Lead can be specified by sheet weight (1 square foot of 1/64-inch sheet lead weighs 1 pound). The type of lead alloy does not have to be specified because the alloy content does not significantly affect the weight or the efficiency of lead as a noise barrier.

Acoustical properties of lead for controlling noise are very good. Qualities such as limpness, mass, and internal damping are essential for decreasing and controlling sound, and lead possesses all three. Lead can be incorporated into other noise-reduction materials and, due to its mass, it can easily be pounded into thin sheets for application—an important design characteristic when maximum space is to be utilized. Mass and limpness should be considered when installing lead or lead-loaded fabric enclosures because these two qualities are most important in sound reduction. Basic considerations and procedures that should be followed for proper installation of enclosures are listed as follows:

1. When installed, sheet lead should not be fastened rigidly to stiffer surface skins because of their tendency to degenerate the limp qualities of the lead.

Table 6-1 Lead Materials for Noise Insulation

Material	Description	Uses
Sheet lead	Usual weight $^1/_2$ lb to 4 lb	Used alone or laminated to substrates of various types
Lead/foam	$^1/_2$-lb and 1-lb sheet lead sandwiched between layers of polyurethane foam	Laminated to enclosures
Leaded plastic	Lead-loaded sheet vinyl or neoprene with or without fabric reinforcement	As a curtain or to line enclosures
Damping tile	Lead-loaded epoxy or urethane tiles	Damping heavy machinery
Casting compounds	Lead-loaded epoxy	Potting, filling complex voids
Troweling and damping compounds	Lead-loaded epoxy and urethane	Damping enclosures, surface resonant members and rattling panels

2. When lead is laminated to another material, its weight should be kept in a composite skin nearly equal to, or in excess of, the weight of the other material.
3. Viscoelastic adhesives or intermittent fasteners should be used when laminating lead panels, rather than continuous fastenings.
4. Leaded panel skins should be used for double-wall construction instead of one skin even when the total weight of both is the same.
5. All seams, doors, and perimeter joints should be caulked or fitted with gaskets to eliminate entrance of any noise.
6. Whenever sound can circumvent a lead shield, all passages should be thoroughly covered. These passages include back-to-back panels, windows, cabinets, electrical outlets, ventilation ducts, the space above a suspended ceiling, and any cracks in walls, however small. All of these passages should be completely blocked with leaded materials.

Sheet lead used for industrial noise control is specified in pounds per square foot. Manufacturers of such sheets make them available in weight from .05 pound per square foot at 1/128-inch thickness to 8 pounds at 1/8-inch thickness. Sheet lead is easily altered to any dimensional need; sheets can be cut with scissors, molded and contoured by hand, and applied to most surfaces with adhesives, as well as laminated to substrates such as aluminum and steel.

Another material that has been used for industrial sound attenuation is sheet lead and polyurethane sandwich material, which can be applied in such places as the inside of machine guards or shrouds. The lead-foam material is available from the manufacturer with 0.5-pound or 1-pound per square foot sheet lead laminated between various thicknesses of polyurethane foam. Thickness of the foam can be obtained anywhere from 0.25 inch to 2 inches in increments of 0.25 inch. Lead-foam material can be obtained with either two or three layers of sheet lead within the laminate. This material may be cut with ordinary scissors or steel rule dies when repetitive parts are required. The material can be anchored down with adhesives and can be shaped and molded.

Another sound control material used by industry has been leaded vinyl sheet, an example of which is two sheets of lead-loaded vinyl laminated to a core of glass fiber cloth to give added strength and durability. This material is specified by pounds per square foot and is available in 1/2, 3/4, 1, 1 1/2, and 3 pound weights. It is commonly applied in the same way as sheet lead and can be used with acoustical wool. Lead loaded epoxy is another type of sound-reducing material. Powdered lead mixed with epoxies makes an excellent damping compound to control structure-borne noise. This substance can be applied by troweling it to surfaces to reduce noise and vibration.

PLENUM BARRIERS

The use of suspended ceilings is very popular in modern office construction. However, this type of ceiling installation allows a space or plenum between it and the previous ceiling from which it is hung. An acoustical problem may result, usually because the material employed in a suspended ceiling is lightweight and does not provide an adequate sound barrier.

A better system has been to use plywood lead sheets, with 2-pound lead, around a frame. These plywood-lead sheets can be covered with different types of wood panels. Office doors can also be made of the same materials. An office with lead-plywood ceiling panels and doors is more efficient in achieving sound attenuation; however, any sound made within the office will go through the ceiling and into the open space above. The noise will then find its way through the ceilings of other offices. The flow of sound through the plenum can be stopped by hanging plenum barriers. Installing such systems does not require excessive materials and is more feasible than laying sheet lead over ceiling tiles.

Plenum barriers are also used for sound attenuation in piping and air-duct systems. In such systems, holes are cut into the barrier, thus leaving an air leak around the pipe or duct that should be caulked or taped. In addition, the pipes and ducts may be lined with lead to lessen the radiated noise. To develop such a noise-attenuation system, the pipes should first be insulated with a paper-backed fiberglass material or mineral wood insulation applied tightly enough to support lead. Next, lead sheets are wrapped around the pipe in sections with each piece overlapping the preceding one by 1.5 inches. The lead sheets are sealed with a standing seam with the overlap taped. For air ducts, a resin-bonded fiberglass board is commonly employed. This material is normally available in 2 × 4 ft boards. The lead is then applied to the ducts in the same manner as for the pipes. Sheet lead plenum barriers are widely used for sound attenuation in office systems. Figure 6-1 illustrates a sound-stopping system employing plenum barriers.

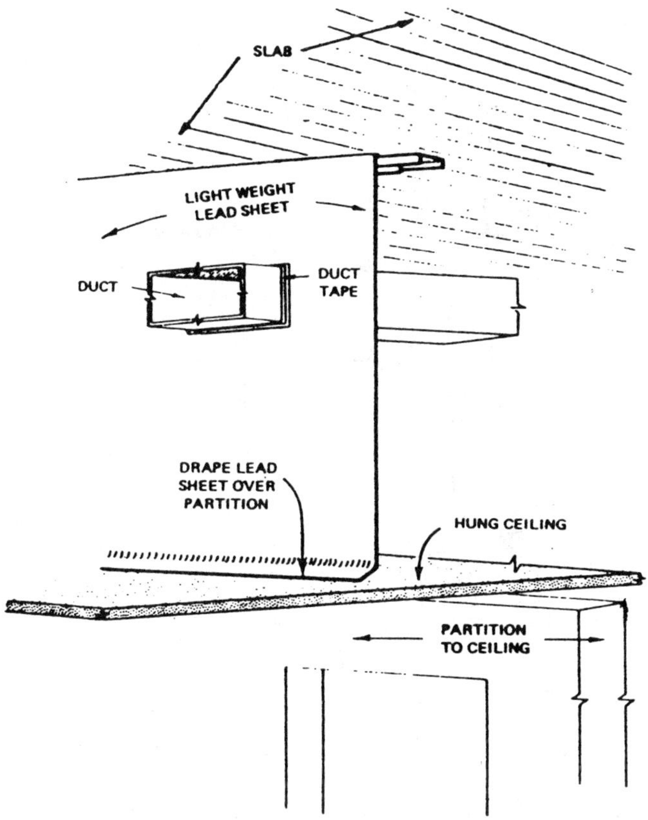

Figure 6-1 Sheet lead plenum barrier system.

HOW TO INSTALL PLENUM BARRIERS

No special tools or skills are needed for installation of plenum barriers. The material is very malleable, 1/64-inch thick, and weighs 1 pound per square foot. Lead will stick with adhesives and can also be taped or cemented into place. Three methods of installing sheet lead plenum barriers are discussed here.

For the first method, begin by cutting the sheet lead 4 inches longer than the total height of the plenum. Then make notches in the top corners about 1.5 × 2 inches in size. Then trim the 2-inch edge over a batten length of metal that is equivalent in length to the width of the sheet minus 3 inches. Lift the sheet with batten and nail on 6-inch centers to the blocking or the soffit of the slab. Finally, hang the bottom end of the lead sheet approximately 2 inches out over the ceiling construction. This method is illustrated in Figure 6-2.

In the second method, first cut the sheet lead 7 inches longer than the total height of the plenum. Make notches at the top two corners 1.5 × 5 inches in area and then turn the 5-inch edge 1.5 times around a wooden batten strip that is 3 inches shorter than the width of the sheet. Now lift the sheet with batten to

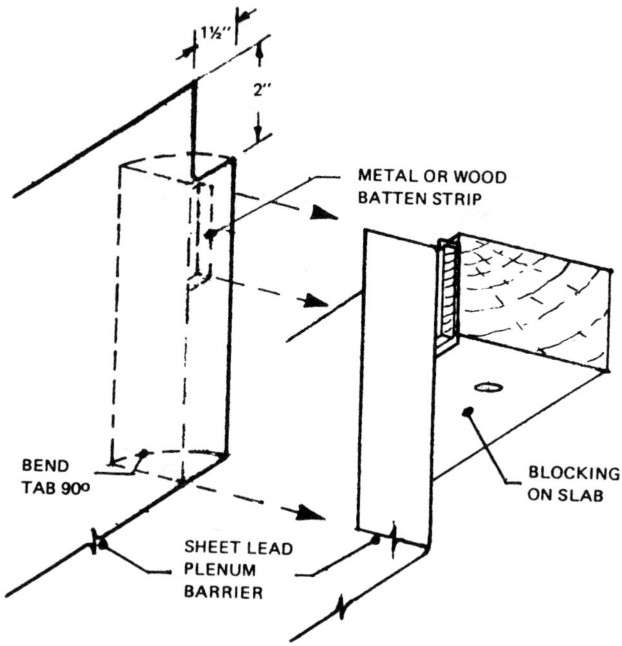

Figure 6-2 Method number one for plenum barrier installation.

the slab and fasten it on 12-inch centers with screws. Finally, hang the bottom 2 inches of the sheet over the ceiling construction. This method is illustrated in Figure 6-3.

In the third method, cut the lead sheet 5 inches longer than the height of the plenum. Then make 1.5 × 3-inch notches in each top corner and turn the 3-inch edge over a 1.5-inch black iron channel that is 3 inches shorter than the width of the sheet. Lift the sheet with channel to the slab and fasten down on 24-inch centers with screws. Then hand the bottom 2 inches of the sheet over the ceiling construction. This method is illustrated in Figure 6-4.

When the plenum barrier is installed, seams can be formed by bending a 1.5-inch tab located at each vertical edge and sealing it with 3-inch tape. This procedure is illustrated in Figure 6-5. For the barrier to give maximum performance in sound attenuation, it must be completely airtight, which can be achieved by gasketing and caulking at all joints.

BARRIERS AROUND PIPES AND WIRES

When installing a plenum barrier, pipes and wires may often interfere. To circumvent such obstacles without sacrificing efficiency of the barrier, the following procedure may be employed. First, cut the lead sheet from the bottom to the point at which the pipe or wire will enter. Next make slits in the sheet to provide for the pipe circumference. Then peel back the segments made by the

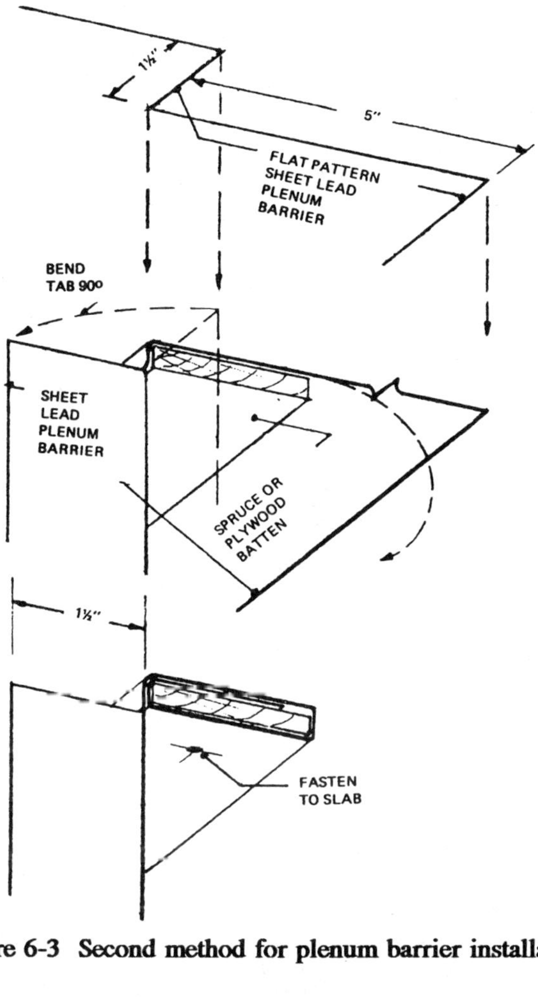

Figure 6-3 Second method for plenum barrier installation.

slits and slide the wire or pipe through the hole. Push the segments back against the wire or pipe so that they are snug against it. Finally, tape around the segments and the slit to the bottom of the sheet. The method is shown in Figures 6-6 through 6-9.

BARRIERS AROUND RECTANGULAR DUCTWORK

To work around rectangular air ducts, the following procedure may be used. Figures 6-10 and 6-11 illustrate the procedure. First, attach adjoining sheets except for making a standing seam at the vertical edges as shown in the figures. Slit the barrier sheets as illustrated, and bend the flaps to fit the duct.

Then tape the duct in place, bend back and tape the flaps. Next, form and tape a standing seam at the vertical edges.

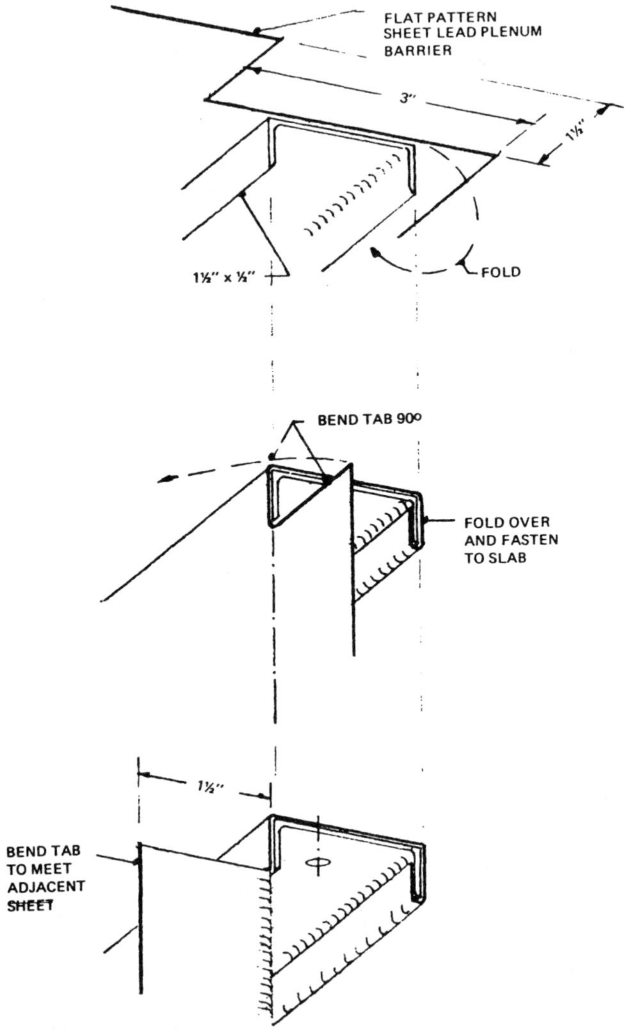

Figure 6-4 Third method for plenum barrier installation.

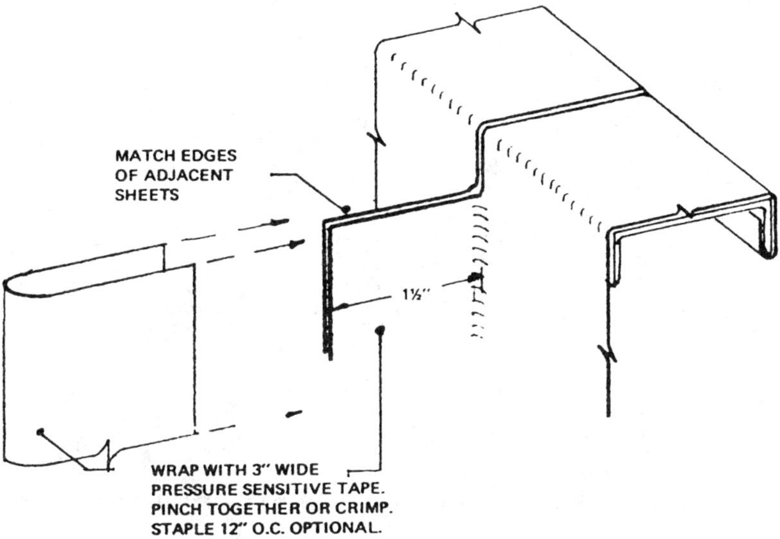

Figure 6-5 Formation of seams when installing plenum barriers.

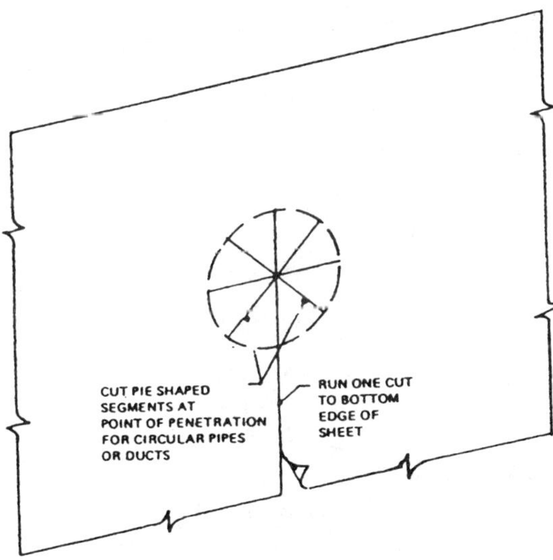

Figure 6-6 Slits made in a plenum barrier to accommodate pipes and wires.

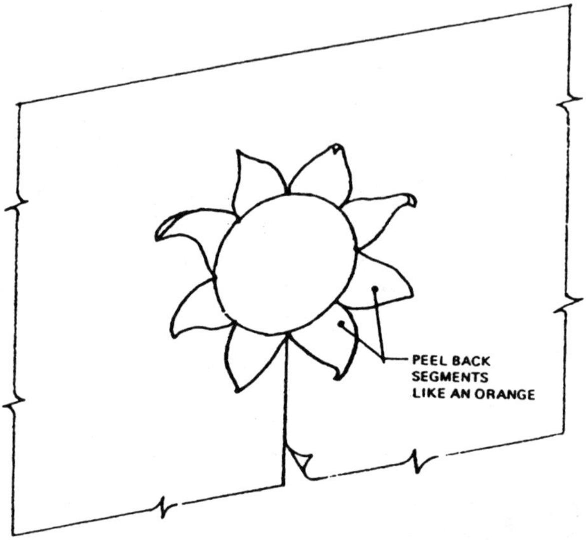

Figure 6-7 Second step in installation of barriers around pipes and wires.

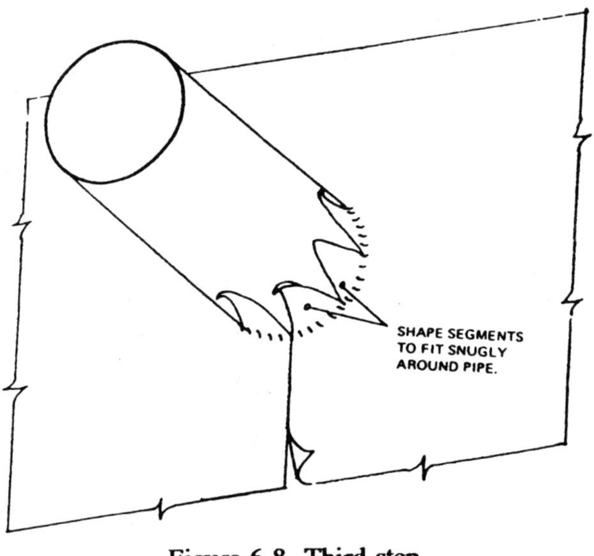

Figure 6-8 Third step.

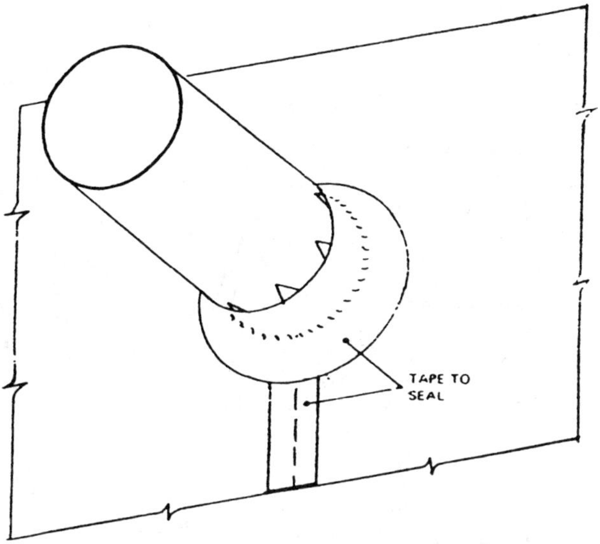

Figure 6-9 Fourth step.

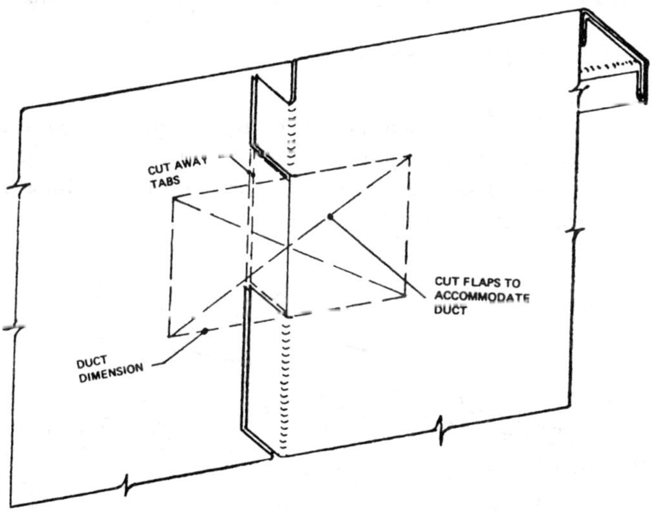

Figure 6-10 Cuts and seams to be made when installing barriers around rectangular ductwork.

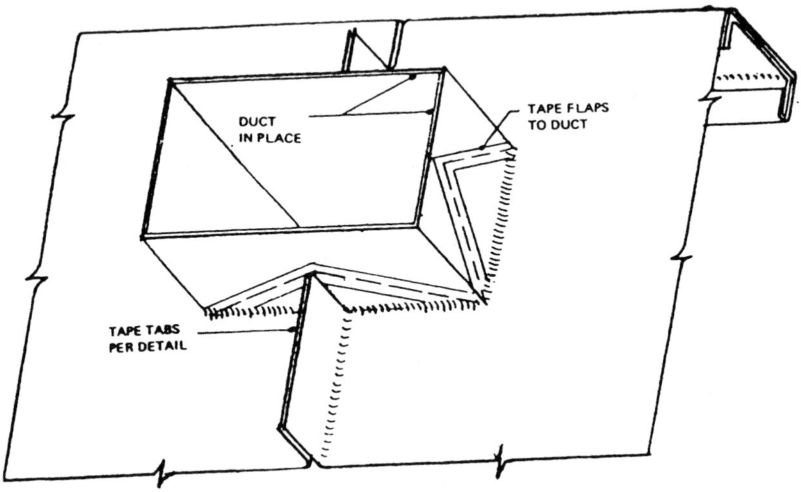

Figure 6-11 Final taping procedure after duct is inserted.

BARRIERS UNDER TWO-WAY JOIST SYSTEMS

To install barriers under pan-type floors, the method illustrated in Figures 6-12 and 6-13 can be employed. First, fasten the lead sheet to the slab. Then cut sections of 1-pound sheet lead to place into the cavities of the floor construction. Finally, tape the bottom edge to the ground.

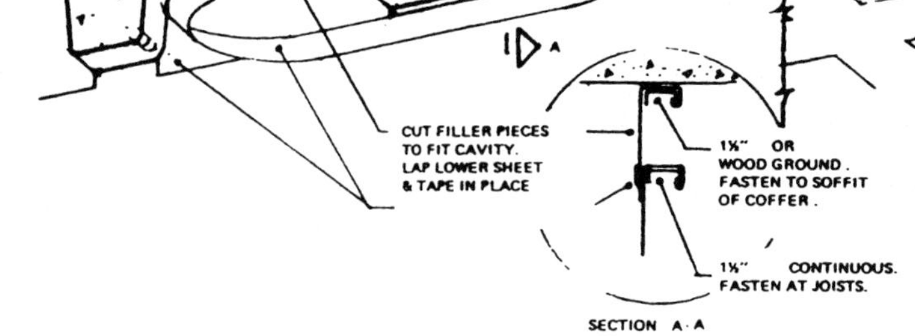

Figure 6-12 First step in installing plenum barriers under two-way joist systems.

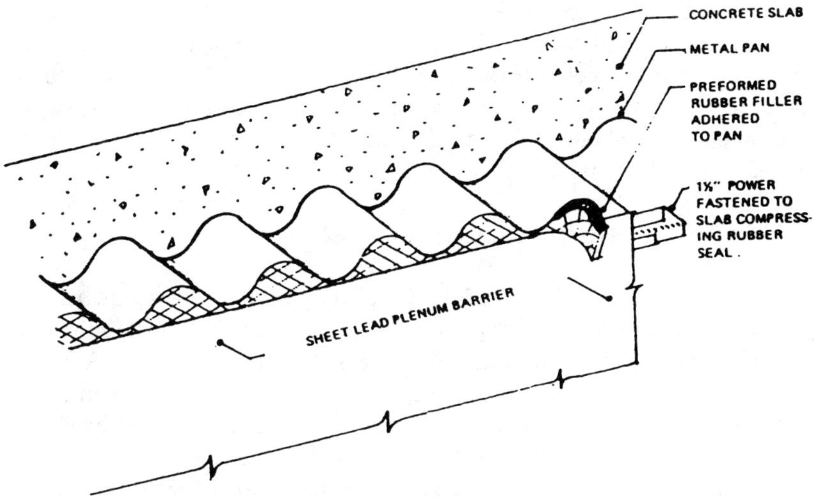

Figure 6-13 Second step.

HIGH TRANSMISSION LOSS CEILINGS

High transmission loss ceilings can attenuate sound which is emanating from upper floors. Such a ceiling is illustrated in Figure 6-14. Two-pound sheet lead

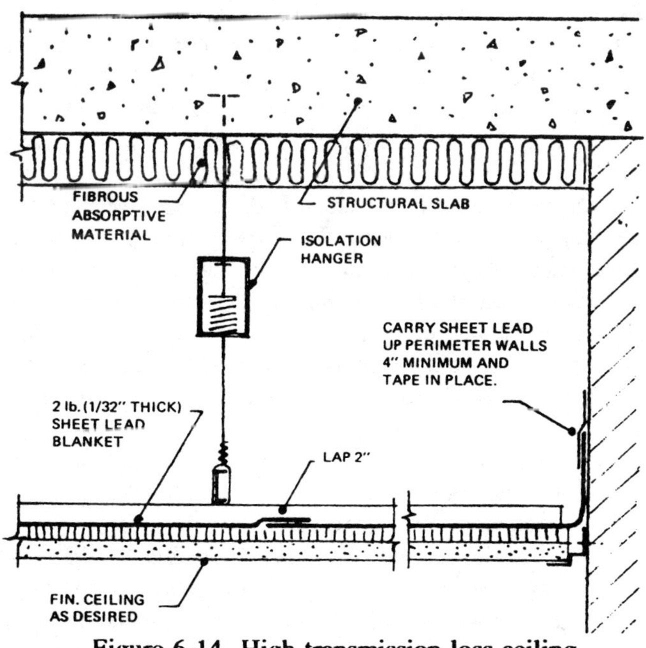

Figure 6-14 High transmission loss ceiling.

lead is normally employed that can be fastened to a board using staples or adhesives before installation. The sheets should be lapped at least 2 inches and carried up the perimeter wall at least 4 inches. The perimeter should be thoroughly caulked to obtain an airtight seal. Also, at each point of suspension isolation hangers should be placed. Last, the ceiling slab above should be covered with a fibrous absorptive material.

ADDITIONAL SOUND-CONTROL MATERIALS

In addition to lead and lead-loaded products, there are many other materials used for sound attenuation and vibration control. Materials with high sound-absorption qualities generally have soft porous surfaces. When sound waves come in contact with these absorptive surfaces, air travels in and out of the pores in the material because of pressure changes produced by the sound. Frictional forces that result from this action convert the sound energy into heat, even though the initial amount of energy is quite small. Every time sound waves travel through the material at each reflection, energy is dissipated, and there is a resulting reduction in the reverberant sound level of the particular enclosure. The use of acoustical material can cut down sound levels in an entire plant as well as limited areas.

ACOUSTICAL FOAMS

Many different types of foams are used widely by industry for sound and vibration isolation. Most industrial foams are available in pore size from 10 ppi to 100 ppi (pores per linear inch). Textures are generally coarse and abrasive in the 10-ppi grades to soft and downy in the 100-ppi grades. Acoustical foams that have excellent high- and low-temperature features can withstand temperatures as high as 250°F, so they can be sterilized with boiling water or steam.

One of the most important factors determining the behavior of a foam sound absorber is its resistance to air flow. In flexible urethane foams, this concept is called *permeability*. Permeability measurements are made by first measuring air flow resistance of the foam and then inverting. For example, a completely closed-cell foam would totally resist air flow and have a zero permeability. As a more open-cell or reticulated foam is achieved, the measurement of permeability rises to 30 percent, 50 percent, 70 percent and so on.

A common misconception concerning flexible urethane foams is the belief that two foams that look alike (have the same thickness, cell size, and density) will provide equal sound absorption. In reality, these two foams may have permeabilities differing by several orders of magnitude and have widely different acoustical properties. To understand this one must understand how sound or noise becomes absorbed. In principle, noise contacts the foam structure in the form of sound pressure waves. The pressure wave within the foam structure is converted to heat energy and is dissipated. The ability of the foam to absorb sound is determined primarily by its permeability and its thickness. Thickness is easily seen and measured, but permeability is neither easily measured nor visually compared.

A very practical, on-the-spot-method of determining whether a foam has permeability is to put it against your lips and blow on it. If it totally resists the pressure you put on it, the foam is not permeable. If you can blow through it, it has a degree of permeability. To understand the reason for the enormous variations of permeability encountered in flexible urethane foams, it is necessary to know something about foam geometry. Urethane foams are formed by mixing reactants that simultaneously polymerize to yield a polymer and generate a gas, which causes the reacting mass to expand into a foam.

Initially these gas nuclei are very small (the number of gas nuclei generated per unit volume determines cell size). However, as the reaction proceeds the gas bubbles grow larger until, at the point where the porosity of the expanding mass has reached a level of about 72 percent (density of about 20 PCF), the gas bubbles become distorted into polyhedrons and the contact points develop into planes. At the instant of completion of expansion, the fully expanded low-density foam consists of nested polyhedrons (ideally these are dodecahedrons having 12 facing planes). Each common face between the polyhedrons contains a thin polymer film or membrane. At this point, the result is a closed-cell foam, completely pneumatic and having zero permeability.

To produce an open-cell foam, it is necessary to design the formulation and processing conditions so that at the exact point the polymer mass is expanded and set into a foam of the desired density, some of the films or membranes separating the cells collapse and establish communication between the cells. Assuming the cells are dodecahedrons, it is necessary for one sixth of the membranes to open to achieve an open-cell foam.

Thus, urethane foams have a strand-and-membrane configuration, the strands defining the edges of the nested polyhedrons and the membranes the planes or faces (often called pores) of the polyhedrons. The permeability of a foam is proportional to the number of ruptured membranes. The membranes comprise only a very small portion of the polymer mass. In the foam industry, the term *level of reticulation* is often used to indicate the percentage of the membranes that are missing. Foam containing no membranes is said to be 100 percent reticulated.

The number of membranes present in a foam is, of course, the major factor in determining its permeability. Expressing permeability as its inverse flow resistance, fully reticulated foams have a flow resistivity ranging from about 20 Rayls per inch for a very fine cell foam to about 1 Rayl per inch for a very large cell foam. However, a foam of any cell size which has most of its membranes intact will exhibit a very high flow resistivity, often greater than 1,000 Rayls per inch.

The conventional bun stock process for urethane foam manufacture presents the foamer with some formidable obstacles. The bun or loaf of foam, perhaps 3 inches thick, emerging from the machine and destined to be cut into sheets has an inherent variation in permeability from top to bottom. Maintaining average permeability involves cutting and testing sections, so rapid feedback for process control is difficult, if not impossible.

The optimum permeability for an absorber depends upon its thickness. As a general rule the thinner the sheet, the higher its flow resistivity to achieve optimum performance. Again, the thin-sheet casting process is advantageous since the permeability of the single sheet being produced can be optimized for its thickness. However, the bun stock manufacturer is faced with the choice of compromising on an average permeability or producing buns having different permeabilities for each sheet thickness. The second choice leads the bun stock manufacturer to the additional serious problem of controlling inventory.

The ability to control permeability in foam casting and thus achieve predictable acoustical results has many advantages for engineers concerned with noise control. Principally, they can now select a foam absorber having good performance in any desired frequency range. These factors can be written into specifications with the assurance that the supplier can deliver a consistently uniform foam absorber for each application.

FOAM WITH POLYMER FILM SURFACE

A foam consisting of an acoustic foam for absorbing sound at peak efficiencies combined with a polymer film surface is an option. The polymer film surface resists scuffing and abrasion, is self-repairing against punctures, and has no tendency to tear. The material is specifically designed for areas demanding a high degree of sound absorption in a visible foam surface. Sheets 24 × 54 inches and 0.5, 0.75, or 1 inch thick come in a hexagonal pattern that can be folded, bent, or worked into complicated configurations without degrading their acoustical efficiency. Its temperature range is from 40°F to 250°F, and it is resistant to petroleum and alkalis. Sound absorption qualities are illustrated in Figure 6-15. Figure 6-16 illustrates a typical application for this product.

ACOUSTICAL PANEL

An acoustical-panel example is a composite of a 5-mil white polyester film chemically bonded to a foam structure that utilizes controlled permeability for optimum acoustical performance. The product also utilizes a pressure-sensitive adhesive backing for ease of application. The film facing, combined with the proper air-flow resistance of foam, yields excellent low-frequency absorption performance. The facing, while being attractive, is durable and well suited for use in environments in which it would be exposed to rough handling, grease, moisture, solvents and other contaminants. The acoustical panel has good low frequency performance. Figure 6-17 illustrates its acoustical qualities at various frequencies.

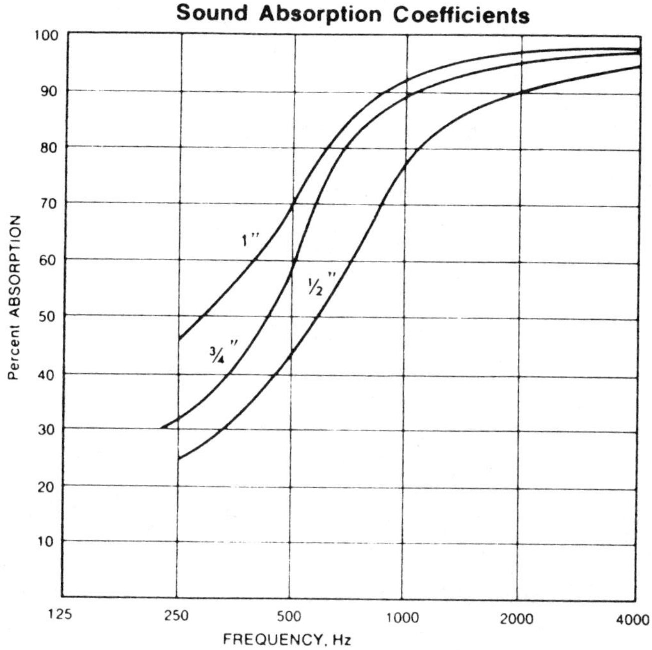

Figure 6-15 Acoustical performance of a foam with a polymer surface.

Figure 6-16 Application for polymer-coated foam.

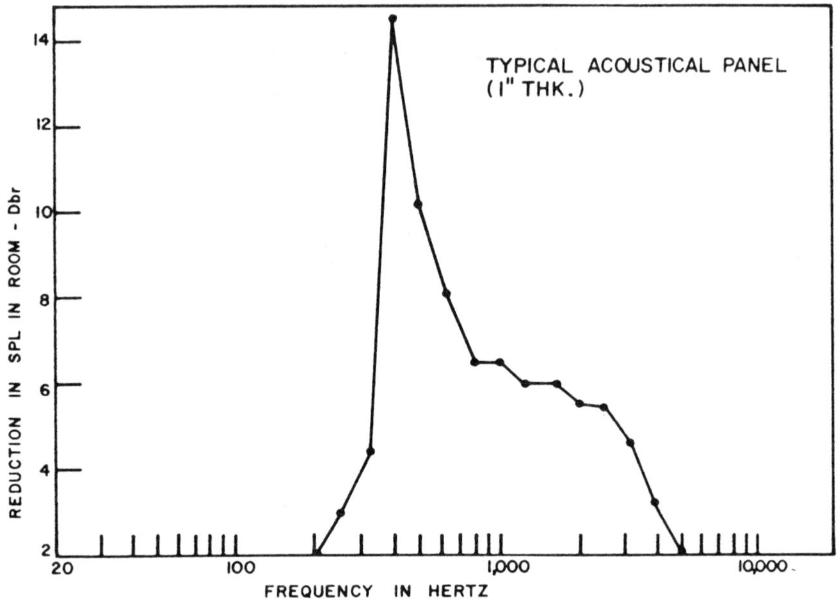

Figure 6-17 Sound qualities of an acoustical panel.

POLYURETHANE FOAM

Flexible polyurethane foam is a highly versatile material that can solve a wide variety of acoustical problems. The number of uses is constantly growing, based on different specific application requirements and unusual foam structure designs that successfully meet them. In the increasingly important area of noise reduction, polyurethane foam has proven to be practical. There is no secret ingredient involved in the sound-absorption successes of polyurethane foam. Actually, what is most important about it is what is missing.

The remaining skeletal strand structure of the fully reticulated open-pore polyurethane foam is able to absorb noise, causing a loss of sound energy by weakening sound waves through reflection.

Foam is a sound-absorption material used for middle- and high-frequency applications in which an attractive appearance is desired, such as hard walls and vehicle headliners. It is also effective as absorbing baffles. Because it can be formed to compound canes with a minimum of wrinkling, it is used for vehicle headliners. It can also be used as absorbing liners for appliances, truck cabs, golf carts, snowmobiles, and office equipment.

For low-frequency applications, the efficiency of foam can be improved by spacing it away from a hard surface. A service life of ten years is expected under normal, in-plant, or vehicle conditions. It will meet random incident absorption coefficient, aging, durability, and flammability requirements. Figure 6-18 and Table 6-2 give data on the absorption coefficients of this material.

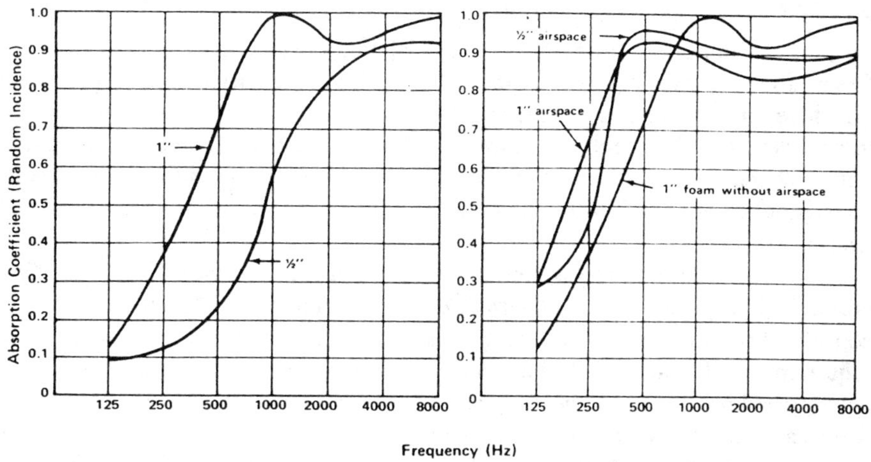

Figure 6-18 Acoustical qualities of polyurethane foam.

Table 6-2 Absorption Coefficients of Polyurethane Foam

Thickness (inches)	Frequency (Hz) 125	250	500	1000	2000	4000	8000	NRC[a]
1/2	0.90	0.11	0.22	0.60	0.88	0.94	0.90	0.45
1	0.13	0.22	0.68	1.00	0.92	0.97	0.99	0.71

[a]Noise-reduction coefficient (NRC) is the average absorption coefficients at 250, 500, 1,000, 2,000 Hz. Coustifoam can be "tuned" to provide maximum NRC by varying its spacing from the mounting surface.

Test procedures outlined in ASTM-C384 investigate the effects of many variables on the ability of flexible polyurethane foam to absorb unwanted sound. These variables include permeability, pore size, thickness, density, stiffness, and surface treatment.

The effect of permeability on the overall absorption capability of foam is interesting. In general, completely closed-cell foam shows minimal capability to absorb sound energy, regardless of thickness. Some open-cell foams have excellent absorptiveness in 1-inch to 2-inch thicknesses. Fully reticulated open-pore foam in thicknesses of 2 inches and greater has broadband absorption capabilities.

In certain noise-control applications it is necessary to protect the foam from its environment. This can be accomplished by laminating a film to the foam, making it more resistant to abrasion and impervious to liquid penetration. However, complete lamination of the film surface to foam gives good absorptive characteristics in only a narrow frequency range. It has been found that a partial lamination considerably improves the sound absorption characteristics of the package.

An important property of industrial component materials is the level of fire retardancy. Many polyurethane foam products are formulated to provide a high level of fire retardancy, and such formulations have been successfully tested in accordance with accepted industry standards. It is also important to note that polyurethane foam can be dyed or painted to achieve specific color effects throughout a wide range without appreciably reducing its ability to absorb sound energy.

In manufacturing sound-absorbing foams, the variables—including permeability, pore size, thickness and surface finish—are taken into careful consideration to produce products that precisely meet specific application requirements. Foams should be selected carefully by industrial designers who have prior knowledge of the frequency range involved, the desired thickness, fire retardancy, appearance/color, cleanability, abrasion resistance, and cost.

Polyurethane foam is used very successfully in reactive mufflers to further reduce noise from pneumatic control valves. A foam disk that is 1 inch thick is placed adjacent to the outlet of the muffler as part of the unit. This application increases the effective noise reduction of the muffler by an additional 9 dBA.

Another unique and interesting muffler design is found in computer housings, where sound waves pass through access holes into a hollow pillow-like polyurethane foam structure. The sound hits two internal baffles and then is directed, right and left, out the ends. In all cases, noise levels have been reduced considerably.

Foam has been used in air-flow acoustical mufflers, especially where space is restricted. Air paths can be designed as eccentric curves within foam blocks, for two reasons: (1) to give the noise a longer dissipation path to travel, and (2) to cause the noise to impinge on a more absorptive foam area, due to the erratic route. Units currently using this concept have reduced noise levels by 8 to 12 dBA.

In a large business office, the noise from a copy machine, vacuum cleaner disturbed employees working nearby. The installation of a 1-inch thick ring of very fine pore reticulated polyurethane foam around the outlet of the vacuum pump reduced the noise level by 9 dBA.

Polyurethane foam was recently designed into several components of a complex plastic pelletizing machine. This involved partial modification of the enclosures for the inlet and outlet chutes, the cutting mechanism, and the feed mechanism. Noise output from the machine was reduced by 13.5 dBA.

Acoustical wall treatment is a new and rapidly growing architectural application for foam. Materials consist of a fine-pore reticulated foam with a fiberglass-woven fabric laminated to it. The treatment is applied like thick wallpaper. At a thickness of 3/8 inch, the covering has an NRC (noise-reduction coefficient) of 0.40, or in effect a 40 percent decrease in noise. At a l-inch thickness, it tests at NRC 0.85, or 85 percent noise reduction.

Another similar application is the use of 1/8-inch foam covered by an open-weave textile material, laminated together in the form of an attractive drape. In the completely stretched flat position, noise hitting this drape achieved NRC 0.55. In the completely closed or pleated position, the drape reached NRC 0.65.

Flexible polyurethane foams have found application as components in sound transmission barriers. For this use the foam is laminated to a lead-loaded vinyl backing and a perforated textile facing so that it resembles a blanket, in what is called *flexible paneling*. The product can be stretched between frames or installed on a wall for modernistic architectural applications, or simply draped, tentlike, over a piece of noisy machinery in an industrial plant. At 250 Hz, using a 0.4 lb/sq ft vinyl, this blanket application reduced noise by 16 dBA. At 250 Hz, using a 0.766 lb/sq ft vinyl, it reduced noise by 21 dBA. At 1,000 Hz the 0.4 lb/sq ft vinyl blanket reduced noise 52 by dBA, and the 0.766 lb/sq ft vinyl model reduced noise by 50 dBA. Not only is this blanket an effective transmission barrier, but the perforated facing makes it an excellent sound absorber as well.

Open-pore reticulated polyurethane foam has many basic composite forms. Its varied capabilities as a highly efficient sound-absorptive material are well documented, and its versatility in application is limited only by the imagination and creativity of the design engineer. In the responsible analysis of noise problems and their solutions and in virtually any engineering application, open-pore polyurethane foam deserves prime consideration.

NYLON IN NOISE REDUCTION

Nylon Gears

The noisiest part of any machine is usually gearing, simply because of the inherent function of transmitting energy from the power source to the production end of the machine. Metal-to-metal gears clang, whine, and otherwise generate noise that is harmful to a worker's ears. Originally selected as a long-wearing alternative to bronze, steel, and phenolic gearing, the noise-reduction properties of a nylon gear have only recently become an important factor in its selection to replace other materials (Figure 6-19).

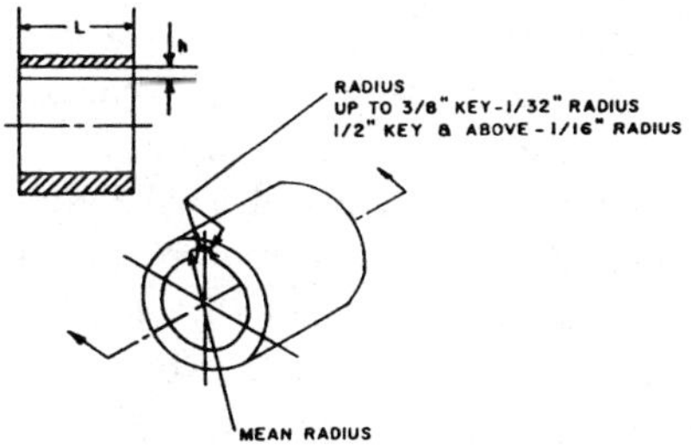

Figure 6-19 Keyway used to drive a nylon gear.

In the past, the paper industry has used more nylon gears than any other industry. They can be found on such machinery as driers, winders, filters, and cutters. Nylon bearings have replaced bronze bearings on overhead cranes.

A large West Coast mill of a major American paper company was experiencing excessive noise from steel gearing on its high-speed continuous rewinder machines. These high-speed machines, used in the manufacture of bathroom tissue and roll towels, generated gear noise that severely limited the amount of time the operators could stay at their stations without the use of ear protectors. In searching for a method of increasing operator productivity without resorting to ear protectors, which are at best a stopgap measure that limits worker effectiveness, the paper mill engineers decided to use nylon for every other gear on the drive train of the rewinding machines. Result: an immediate measurable reduction in the noise level.

PVC FILM

Polyvinylchloride is another material that has been used by industry for noise attenuation. In the form of a transparent noise barrier material that is limp and flexible, PVC film resists the passage of sound waves and reduces noise transmission. It permits machine operators, for instance, to monitor their machine even though it may be surrounded by an acoustical barrier, such as a curtain with one or all panels made of this material. It reduces noise by an average of 20 dB (STC). Made of tough, resistant cast PVC, it is resistant to yellowing, fading, and clouding in normal industrial applications. PVC film is available in 20-yard rolls, 54 inches wide, 0.070 inches thick, and weighing 0.5 lb/sq ft. This material is recommended for use as hanging curtains and as "windows" in curtains and fabricated enclosures for noise control systems where visual monitoring is desired.

Typical enclosure-type applications include: presses, machine tools, textile looms, twist frames, conveyor lines, packaging lines, office equipment, and control panels. It can be used alone or as window panels in opaque noise control systems.

Fiber-loaded vinyl is a flexible noise barrier material that is a limp, tough, high-temperature-fused vinyl, loaded with high-density, nonlead fillers and reinforced with high-strength fabric. It resists the passage of sound waves and reduces noise transmission because it doesn't resonate the way metal mass barriers do. Available in standard 0.5 and 1 lb/sq ft densities with acoustical ratings of STC 20 and 27, respectively, it can be used in layers to achieve higher STC levels and follow the limp-mass loss principles of noise transmission. This material has high tear and tensile strengths; it will not rot, shrink, or cause metal corrosion. And it is easy to fabricate and install because it can be cut with a knife or scissors, or die-cut. Fiber-loaded vinyl is also safe and nontoxic. It is available in a variety of colors, with textured or embossed surfaces, with Saran coating for USDA-approved installations in food processing plants. This material has been confirmed by plant engineers, architects, and design engineers to be an effective noise barrier in industrial, OEM, new construction, and remedial applications such as rollaway curtains, pipe- and duct-noise lagging, machinery covers, ceiling noise barriers, crosstalk barriers, wall and door septums, noise panels, and rooftop-equipment barriers.

There are many lead-loaded vinyls employed by industry for sound control. Lead-loaded vinyl fabric is a fiberglass-reinforced material coated with lead-loaded vinyl for the reduction of noise transmission, ranging from 20 to 29 dB (STC). Higher STC levels can be achieved with multilayers of this material. Made with vinyl for toughness, limpness, durability, appearance, and oil and chemical resistance, it is reinforced with fiberglass for strength, durability, and flame resistance. It is loaded with lead powder for weight (limp mass) and maximum noise attenuation with minimum thickness.

Lead-loaded vinyl is available in many industrial and aircraft grades, with densities ranging from 0.14 to 1.50 lb/sq ft. It is also used with polyurethane foam for improved sound transmission loss, and with aluminum foil to provide reflective thermal insulation and a noise barrier in one material for pipes and ducts.

It is easy to fabricate and install because it can be cut with a knife or scissors, or die-cut. Approved by the Federal Drug Administration for use in food and drug plants, the fabrics reduce airborne and structure-borne noise. It can be used alone, in layers, or in combination with acoustical glass, wool, or sound-absorption foam. Typical applications for the use of this product include: aircraft sound blankets, gearbox pillows, air-intake-duct bellows, insulated pipeline lagging, door and panel barrier diaphragms, noise-isolation enclosures, and impact or sympathetic vibration noise dampers. It is not recommended for dynamic movable curtain applications requiring USDA approval or air-handling ducts and plenums where building codes require UL 181 air-handling specifications.

DAMPING COMPOUND

There are also compounds used for reducing excessive noise and vibration that meet noise-control levels under the Occupational Safety and Health Act (OSHA 1910.95 and others). Damping compound is highly effective for the control of noise caused by the impact of one material upon another (for example, steel on steel) and for damping sound generated by vibrating metals and nonmetallic plates or panels. It is easy and fast to use because it can be applied by brush or putty knife directly onto existing equipment. In some cases this work can be done without shutting down production. Another application is the lining of feed hoppers used by large automotive companies. Castings vary in weight from 1 to 10 pounds each, with approximately 2,000 pounds of castings dumped in a series of hoppers three to five times each hour, 24 hours a day, or 7 days per week. Damping compounds in paste and sheet form are in wide use for reducing noise and vibration on sheet metal and similar thin materials.

SILENCERS AND SUPPRESSOR

Of the variety of silencers industry employs for noise attenuation, one is a duct silencer designed to accommodate air flow in either direction of the duct. Figure 6-20 illustrates a typical duct silencer, the top and bottom of which are

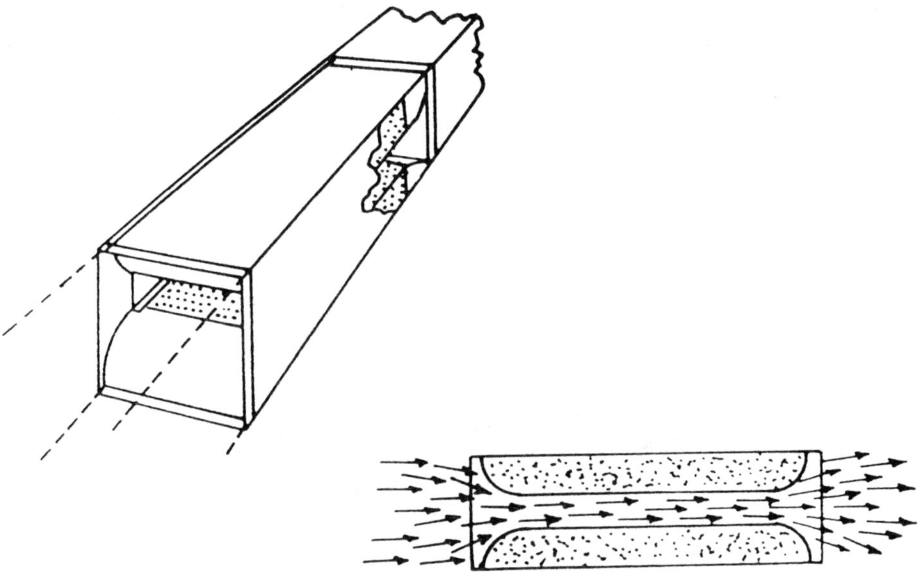

Figure 6-20 Typical duct silencer.

are filled with a sound-absorbing inorganic material and covered with perforated galvanized steel sheeting. Figure 6-21 illustrates a circular duct silencer, which is filled on the inside with an inorganic material along the exterior walls and the inner shaft; this material is covered with galvanized steel.

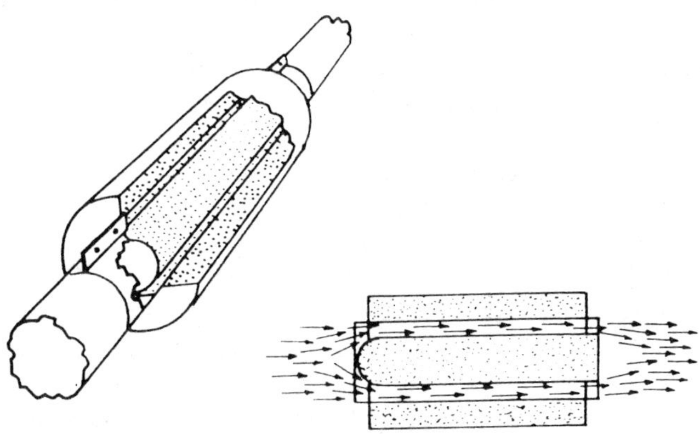

Figure 6-21 Circular duct silencer.

Compact silencers are shown in Figures 6-22 through 6-25. Figure 6-22 shows how the packaged equipment must be lined to eliminate transmission of noise through vibration. In Figure 6-23 we see a compact silencer used between transfer grilles to cut down crosstalk and room-to-room noise transfer. Figure 6-24 shows a compact silencer used in return-air ducts behind a return grille to reduce equipment noise. Figure 6-25 illustrates a compact silencer applied in a suspended acoustical ceiling, with troffers supplying air to a given area. The space above the suspended ceiling is used for return air. This space has long been a problem because of the cross transfer of noise from one area even though the areas have been partitioned off for privacy.

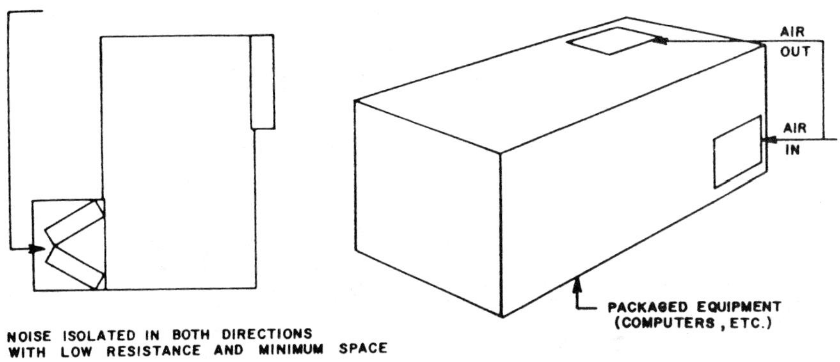

Figure 6-22 Lining packaged equipment stops vibration noise.

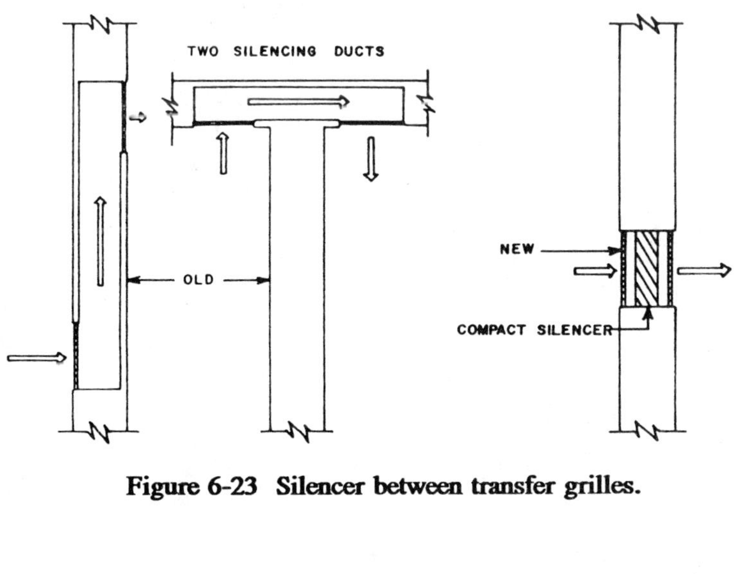

Figure 6-23 Silencer between transfer grilles.

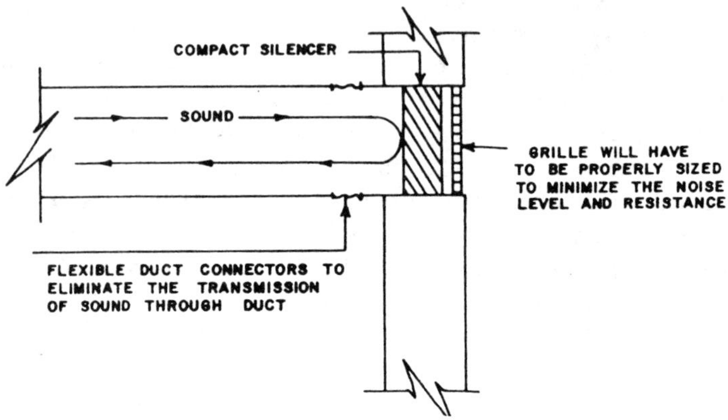

Figure 6-24 Silencer used behind return grille.

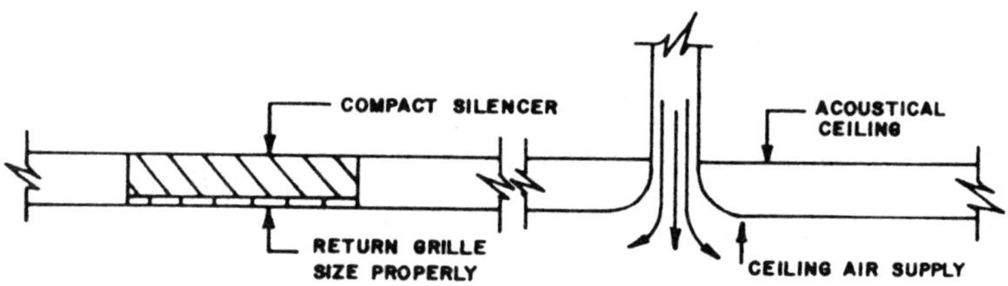

Figure 6-25 Silencer used in a suspended ceiling.

7 VIBRATION

NATURE OF VIBRATION

Vibration may be considered the oscillating motion of a particle or body about a reference position. This motion can be periodic, random, or transient. Simple examples of each are discussed.

Periodic Functions

The sinusoidal signal of Figure 7-1 is by definition at one discrete frequency, which is given by $1/T$, and has the units of Hz (Hertz), which is the number of cycles/second. T is the time taken for the wave to perform one complete cycle. This motion is periodic in that it repeats itself at regular intervals of time (T). The nearest that one would come to a pure tone in vibration is the movement of the arms of a tuning fork.

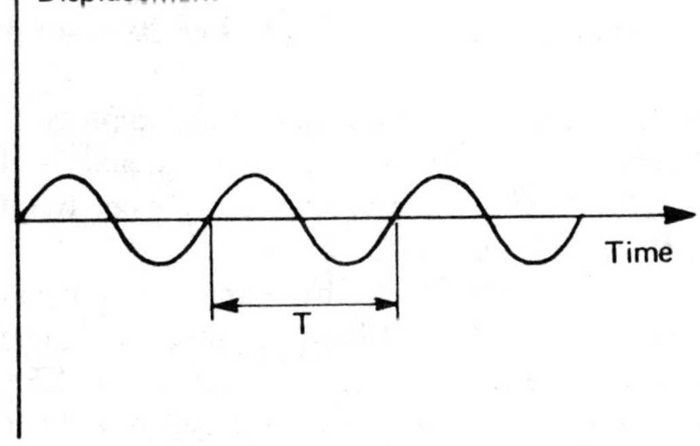

Figure 7-1 Example of a simple harmonic (sinusoidal) vibration signal ($f = 1/T$).

The vibration of Figure 7-2 is still periodic because it repeats itself at regular intervals, but it is now complex (that is, it is not purely sinusoidal). In fact this particular example is a combination of two sine waves and is the sort of vibration produced during the piston acceleration of an internal combustion engine. A periodic signal need not be symmetrical, and it can be made up of many combinations of frequencies of different amplitudes and phase.

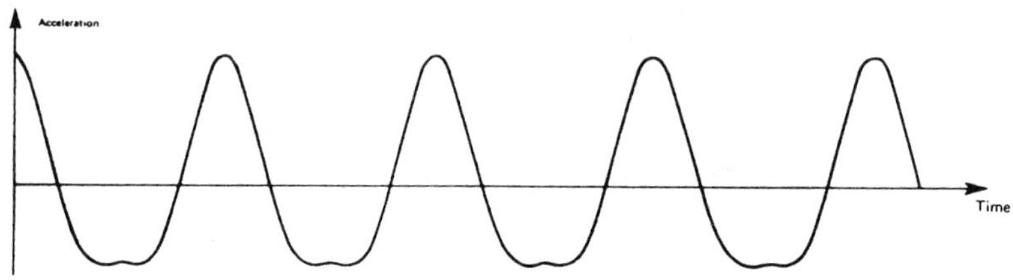

Figure 7-2 Example of compound harmonic periodic motion ($f + 2f$).

The signal in Figure 7-2 is made up of two components of different frequencies and amplitudes, and these are shown in Figure 7-3. In this example one sine wave has a frequency that is twice the other, and its amplitude is much smaller. The amplitudes are summed arithmetically at each point in time to obtain the combined signal level.

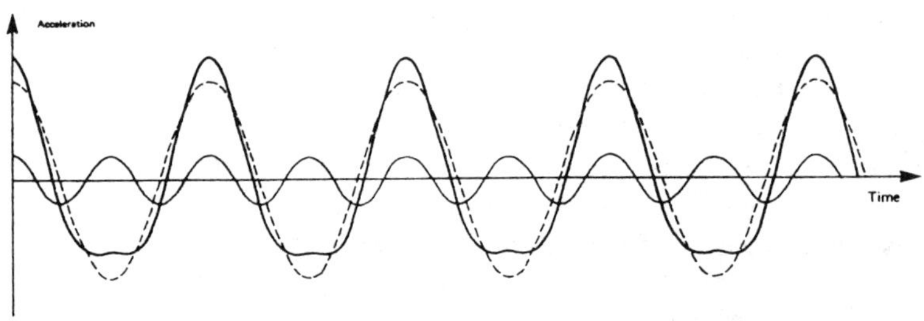

Figure 7-3 Waveform of Figure 7-2 split into its components.

An important and useful way of portraying the amplitudes and frequencies of all the different components of a complex signal is the frequency spectrum, which is a plot of amplitude against component frequency. A few simple examples are given in Figure 7-4.

As can be seen from Figure 7-4A, the simple sine wave has a spectrum of one discrete frequency ($f = 1/T$), and is represented by a line whose height is proportional to the amplitude of the wave. The waveform of Figures 7-2 and 7-3 has two components, each having a discrete frequency. One frequency is twice the other. The two frequencies each have one line on the spectrum, whose height is proportional to amplitude (Figure 7-4B). In this case, the frequency f would be called the fundamental frequency and the frequency $2f$ is its second harmonic, since it is twice the frequency of the fundamental frequency. A "square" wave, shown in Figure 7-4C, is another example of a periodic

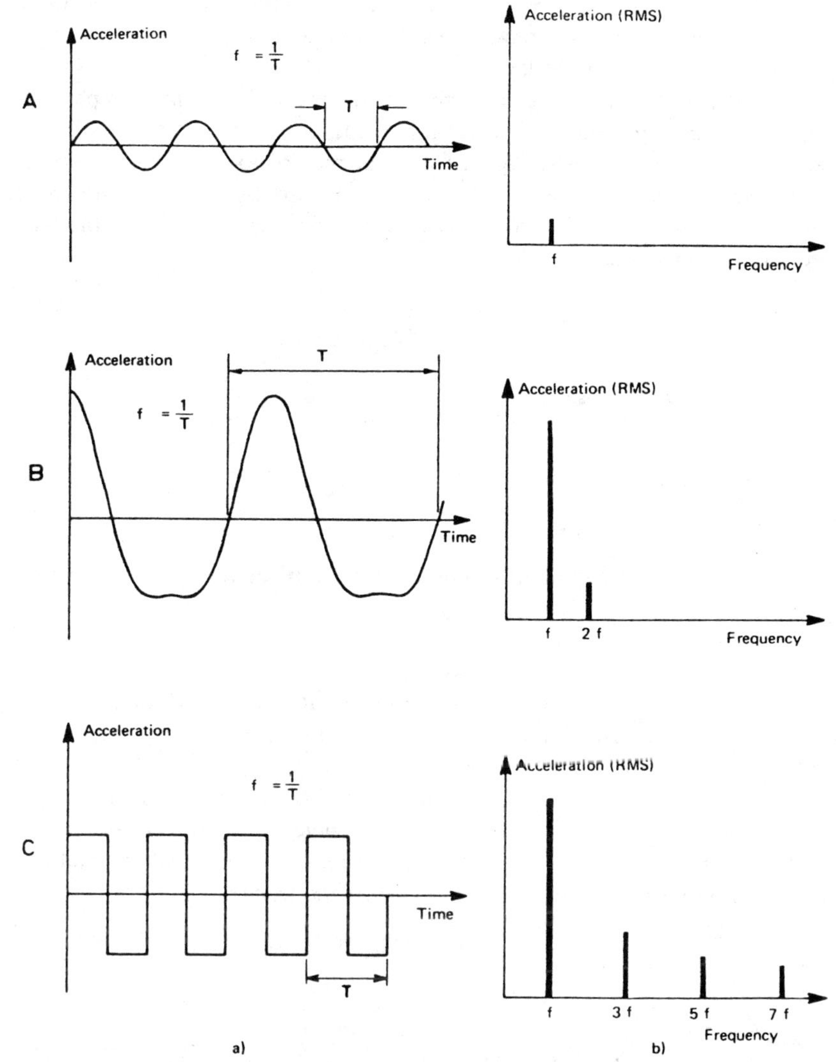

Figure 7-4 Examples of (a) periodic signals and (b) their frequency spectra.

function. This consists of a fundamental frequency f and odd harmonics only (that is, 3, 5, 7f) that drop off in level at a defined rate.

All periodic functions such as those shown previously can be defined by precise mathematical equations, which can make their analysis much simpler.

Random Vibration

The most commonly encountered type of vibration in everyday life is random vibration. It is continuous, but nonperiodic, and contains many frequency components. While many of these components will be related in the form of harmonics of certain frequencies due to machinery movements, many components will be entirely independent of these.

A random vibration signal is shown in Figure 7-5. It lacks repetition, its frequency spectrum is broadband or continuous, since it contains all frequencies, and it has some spikes due to resonance or to harmonics of certain vibrating components. These spikes are often caused by the main source of vibration, and once they are discovered from frequency analysis they are relatively easy to isolate.

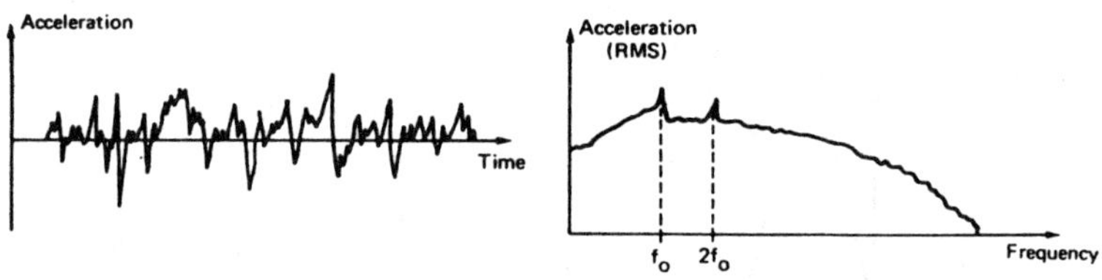

Figure 7-5 Example of random vibration and a typical frequency spectrum.

Transient Vibration

A transient vibration is noncontinuous; it occurs due to impact, or during the starting up of a motor, or anywhere that the exciting force is not continuous. A transient vibration is often deterministic.

The transient vibration shown in Figure 7-6 has a spectrum with many lobes that would contain a whole range of frequencies. The size of the lobes (in terms of frequency range) will depend on the duration of the transient, and their relative amplitudes will depend on the shape of the transient itself.

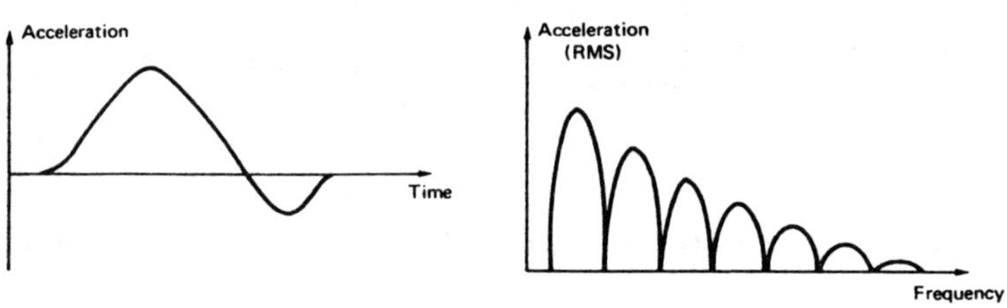

Figure 7-6 Example of a transient vibration and a typical frequency spectrum.

SOURCES OF VIBRATION

Vibration results from some part of moving machinery being out of balance, turbulent fluid flow, rattling of loose objects, impulses, shocks—the list is endless. Vibration is normally undesirable, but it can often be reduced by careful design or by development modifications. Sometimes, however, vibration is introduced to make things work, such as vibrating conveyor belts, mechanical hammers, and even musical instruments. In such cases resulting noise is unavoidable. It is then important to isolate it as much as possible, even to the extent of preventing your hi-fi loudspeakers from rattling your neighbor's picture on your adjoining apartment wall.

VIBRATION EFFECTS

The effects of vibration are often serious. Humans subjected to vibration can be affected by blurred vision, loss of balance, and consequent lack of ability to do their job properly. In some cases, certain frequencies and levels of vibration can permanently damage internal body organs. Machinery can also be damaged by vibration. If the vibration occurs at the resonance frequency of some component, it can be cracked or broken by fatigue, and nuts, bolts, and rivets can be shaken apart. Noise resulting from vibration is also often a serious problem and can be a health hazard to people exposed to it for long periods.

One real difficulty with vibration is that it will not stay in one place unless special steps are taken to isolate it. Vibration is transmitted through any solid object in contact with it, including the floor, walls, pipes, electrical conduits, and any other mechanical linkages, which in turn causes things in contact with them to vibrate and radiate noise. Even if a particular frequency from one machine does not cause any of its own components to resonate, it could quite possibly excite a resonance in a connected machine. It is thus important to measure and control vibration.

VIBRATION MEASUREMENT

It is necessary to measure vibration for several reasons. While sometimes the vibration picture is obvious, there are many occasions when it is not, and it would be senseless to design a means of reducing a harmonic component frequency while vibration continues at the fundamental frequency. From the design point of view, the magnitude and frequency of the vibration need to be known to ensure that the stresses induced are not too great for the material to withstand. Another important reason is that if a certain piece of machinery has been found to resonate at a given frequency, the operator can avoid running the machine at a speed that will excite that resonance. For vibration damping or isolating, it is necessary to know the amplitudes and frequencies involved in order to select the correct damping materials. Preventive maintenance is another area where vibration monitoring is useful, since many faults (like worn teeth on a gear wheel or the start of a roller bearing's failure) can be detected long before failure merely by noticing changes in the vibration spectra.

There are three main parameters used to describe vibration: displacement, velocity, and acceleration. Displacement is the distance moved by the measuring point from its natural position. Velocity is the speed at which that point moves, and acceleration is the rate of change of its speed with time.

Displacement is proportional to strain in the material. Velocity is a function of kinetic energy that must be dissipated in the machine. Therefore, velocity is related to potential damage in the machine.

Acceleration is proportional to the force acting on an object. These parameters are measured in the normal units of distance, velocity, and acceleration, which are meters (m), meters per second (m/s), and meters per second squared (m/s^2), respectively. The units are in accordance with ISOR 1,000 (SI units). Acceleration is also often measured in terms of gravitational constant (g), and it is common to see vibration ranges or levels quoted in terms of numbers of g because it is easy to calibrate measuring equipment at levels of 1 g, and also because g is an internationally understood symbol, regardless of units.

Historically, displacement was the first parameter to be measured because with slow-moving machines and large displacements it was the easiest to see and was measurable by simple optical methods. Displacement limits were then set by the law of elasticity, the fatigue of materials, or by mechanical clearances. Measurement of displacement emphasizes very low frequencies. At higher frequencies of vibration, displacement is negligible.

With the introduction of higher-speed machines, the displacements became smaller and more difficult to see, but breakdown could still occur, so a move toward velocity measurements was made. Many rotating machines have a vibration frequency spectrum with a fairly even distribution of energy across the frequency range up to 1 kHz (Figure 7-7), thus making it reasonably easy to

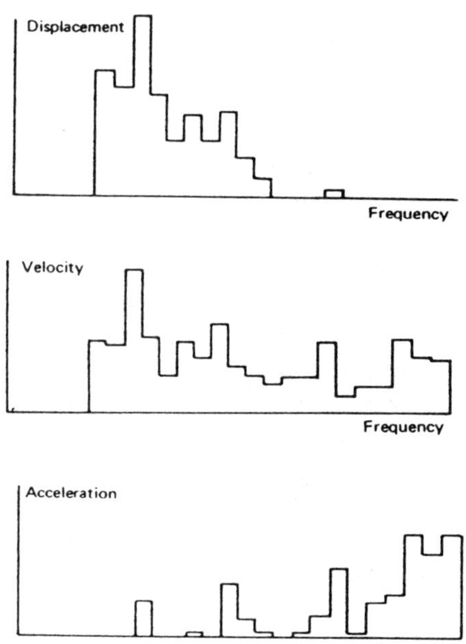

Figure 7-7 Spectra showing the effects of measuring displacement, velocity, and acceleration on rotating machinery in the frequency range up to 1 kHz.

to prescribe a velocity limit for a particular type of machine. (It is even easier to prescribe for single resonance frequencies). Standards organizations in several countries now recommend this type of limit especially for installed electric motors. Measurement of acceleration puts emphasis on the higher frequencies, where it is often important to understand the vibratory forces acting on the machine.

For high-frequency vibration (above about 1 kHz), it is necessary to measure acceleration because displacement velocity drops off rapidly at high frequencies. The reason for this will be seen later. Since high frequencies play a large part in preventive maintenance, early detection of breakdowns and vibration monitoring acceleration measurements are very important. For example, many faults in machine elements, like roller bearings, can be detected in the frequency range 20 kHz to 50 kHz long before failure occurs, thus making it possible to plan repairs and shutdown times.

DISPLACEMENT, VELOCITY, AND ACCELERATION

There is a well-defined relationship between displacement, velocity, and acceleration. The concept of simple harmonic motion serves as a basis for discussing the mathematical relationships among these parameters.

Figure 7-8 shows a wheel rotating at a constant angular velocity of ω radian/s; θ is measured counterclockwise from the bottom, and x is measured to

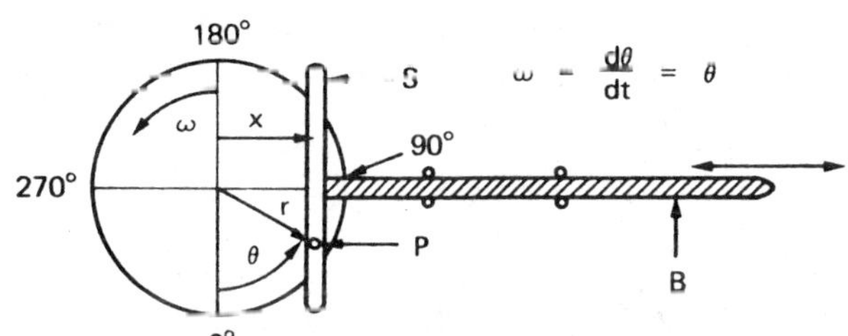

Figure 7-8 Example of simple harmonic motion.

to the right. (There are 27 π radians in one revolution, so the wheel is turning at $\omega/2\pi$ rev/s, or $\omega/120\pi$ rev/min.) The wheel drives a reciprocating mechanism *B* by means of a protruding stud *P*, which is free to move in the slider *S*. *B* is said to move backward and forward with simple harmonic motion (provided there is no mass or friction in any of the sliding contacts) in the horizontal plane. Thus, the displacement (x) of *B* will be as shown on the curve, and one cycle will be completed with each complete revolution of the wheel. In this case revolutions/second or cycles/second or Hertz, and the standard relationship between angular speed ω and frequency f is obtained:

$$\omega = \pi f$$

The displacement x of B from its central position is given by:

$$x = r \sin \omega t$$

when θ = 0° or 180°, x = 0; θ = 90°, x = + r; θ = 270°, x = -r. The maximum displacement is given when sin ϖt = ±1, and is thus equal to r.

Differentiating displacement with respect to time, we obtain the velocity v of b:

$$v = \frac{dx}{dt} = x = r \omega \, cost \; \omega t$$

when θ = 90° or 270°, x = 0; θ = 0°, θ = 0°, x = +rϖ; θ = 180°, x = -rϖ. The maximum velocity, V, occurs when cos ϖt = ±1, and V = r⊃.

Differentiating again with respect to time, we obtain the acceleration a of B:

$$a = \frac{dv}{dt} = \frac{d^2x}{dt^2} = \ddot{x} = -r\omega^2 \sin \omega t$$

when θ = 0° or 180°, x = 0; θ = 90°, x = $-r\varpi^2$; θ = 270°, x = $+r\varpi^2$. The maximum acceleration, A, occurs when sin ϖt = ±, and A = $r\varpi^2$. Figure 7-9 illustrates the effects of these relationships.

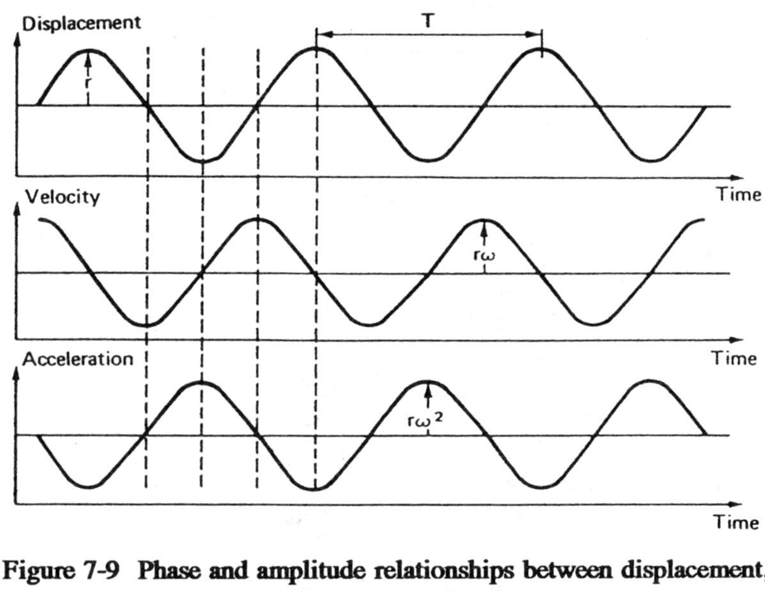

Figure 7-9 Phase and amplitude relationships between displacement, velocity, and acceleration.

It can be seen from Figure 7-9 that there is a phase difference between these components. The velocity leads the displacement by 90° (or one quarter of a cycle), the acceleration leads the velocity by 90°, and so the acceleration leads the displacement by 180°. However, in terms of measurement this phase difference does not matter because RMS or peak readings are usually obtained on a meter or recorder.

It can also be seen that each differentiation has multiplied the signal by ω , which does have practical significance. It means that at higher frequencies, the acceleration signal is the largest by a factor of ω or ω^2. Since $\omega = 2\pi f$, the acceleration signal is often the easiest one to measure, or sometimes the only one that can be measured because the other signals drop into instrumentation noise.

The main significance of the relationship between these parameters is that at any given frequency one measurement can provide the other two. Sensors having a flat acceleration response (accelerometers) are the most common vibration sensors. Integrating the acceleration will give velocity, and integrating again will give displacement, and this is a fairly standard procedure. This integration is performed by use of a low pass filter with a roll-off of $1/f$ for velocity and $1/f^2$ for acceleration.

The emphasis of different parts of the spectrum can be seen from the spectra of Figure 7-7. In this case we have a fairly flat velocity spectrum. The division by ω to obtain displacement attenuates high frequency signals (at high frequencies ω is large), and in this case the signal disappears. The multiplication by ω to obtain acceleration emphasizes high-frequency signals. Thus, while acceleration may not be the parameter we wish to study at high frequencies, it is the one we would measure in order to obtain velocity.

When drawing spectra, we normally plot the frequency on a logarithmic scale, as it has the effect of expanding the lower frequencies and compressing the higher frequencies, and allowing us to have a reasonable resolution without large sheets of graph paper. Logarithmic plotting also ties in well with the concepts of an octave (doubling of frequency) and a decade (multiplying the frequency by 10) because one octave is now a constant width on the paper, as is one decade, regardless of the frequency range it is covering. Linear scales are useful sometimes, particularly for sorting out harmonics.

We also commonly use a logarithmic scale to plot vibration amplitudes. This gives us the decibel (dB) scale, which is also commonly used in acoustics, electronics, and control theory. Part of the use of the two logarithmic scales of level and frequency can be seen in Figure 7-10 where all the parameters now have straight line relationships with each other for one constant parameter. In this case (Figure 7-10) we have a flat acceleration spectrum, and the velocity and displacement drop off with frequency at rates of 6 dB per octave (20 dB per decade) and 12 dB per octave (40 dB per decade), respectively. The displacement drops into instrumentation noise very rapidly, and the velocity signal a little later. At very low frequencies, distortion and cut-off is introduced by the limitations of the measuring and integrating equipment.

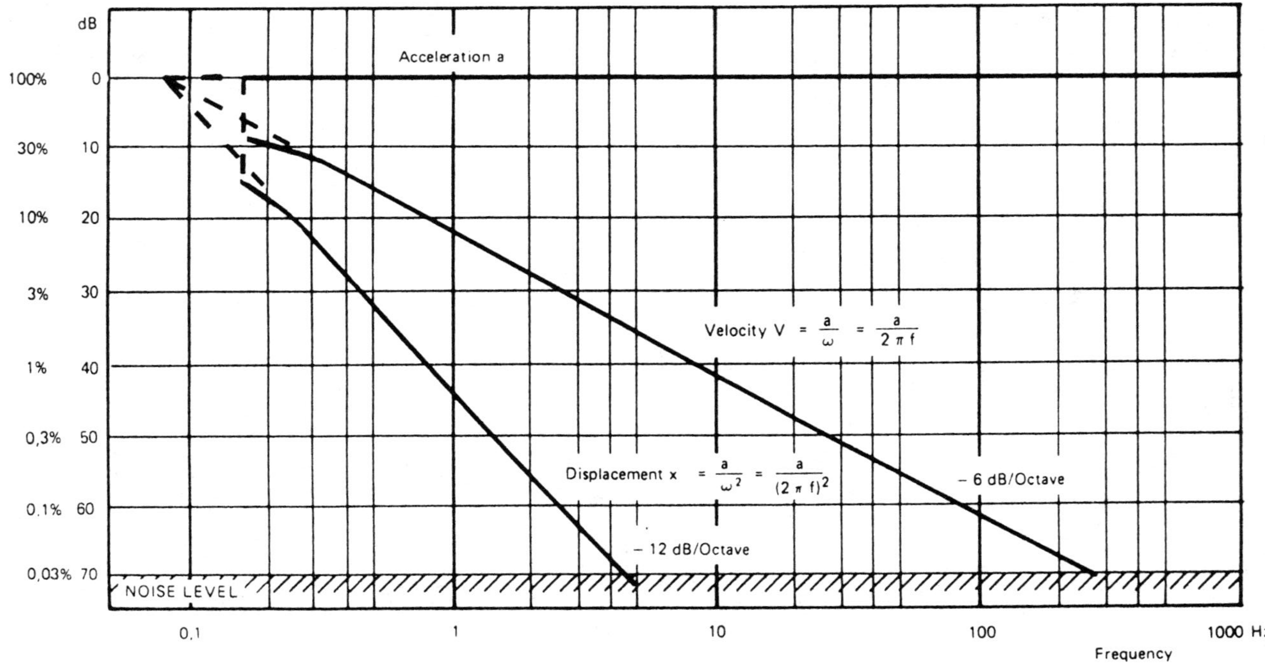

Figure 7-10 Relationship of levels of displacement, velocity and acceleration.

PEAK, AVERAGE, AND RMS VALUES

There are several different ways of quantifying the level of vibration. The first is to use the peak (or maximum) value, as was done with the analogy of a turning wheel earlier (Figure 7-8). This peak value is shown in Figure 7-11.

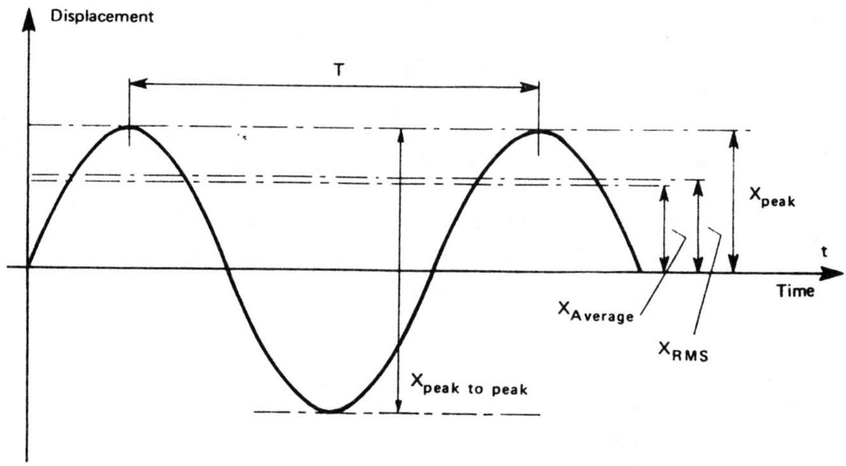

Figure 7-11 Peak, average, and RMS values for a sine wave

It is useful for simple harmonic vibration (such as the one shown), but for other types it is not as useful because it depends only on an instantaneous vibration magnitude, and takes no account of time history producing it.

The peak-to-peak value, or the magnitude of the positive and negative extremes of the motion, is also commonly used. For a symmetrical signal, it is twice the peak value. Another quantity, which does take into account the time history, is the average absolute value, which is defined as

$$x_{|average|} = \frac{1}{T} \int_0^T |x| \, dt$$

Even though this quantity takes into account the time history over a period T, it has been found to be of limited practical interest because it has no direct relationship to any useful physical quantity. A much more useful descriptive quantity, also taking the time history into account, is the RMS (root mean square) value, defined as

$$x_{RMS} = \sqrt{\frac{1}{T_0}\int^{T} x^2 (t)dt}$$

The major reason for the importance of the RMS value is its direct relationship to the energy content of the vibrations.

The relationship between these values can be expressed as

$$x_{RMS} = F_f x|average| = \frac{1}{Fc} x_{peak}$$

where F_f and Fc are called *form factor* and *crest factor*, respectively, and give some idea of the waveshape of the vibrations being studied.

For a sine wave, this will be

$$x_{RMS} = \frac{\pi}{2\sqrt{2}} x_{|average|} = \frac{1}{\sqrt{2}} x_{peak}$$

$$motion,\ F_f = \frac{\pi}{2\sqrt{2}} = 1.11\ (\approx 1\ dB)\ and\ F_c = \sqrt{2} = 1.414\ (= 3\ dB)$$

Most vibrations encountered will not be pure harmonic waveforms, and in general an RMS measurement is preferred.

PRINCIPLES OF VIBRATION TRANSDUCERS

It is possible to measure the acceleration, velocity, or displacement of a vibration.

Displacement Measurement

To measure displacement, or the distance moved by the measuring position on the vibrating object from its natural position, mechanical, electrical, and optical transducers are available. Examples are the vibrograph and the strain gauge.

Vibrograph

An example of the mechanical type of transducer is the vibrograph (Figure 7-12). It is essentially a hand-held box, with a probe connected to a series of levers to magnify the movement and to record the vibration onto a moving trace. The trace can be enlarged optically, and the displacement and frequency measured.

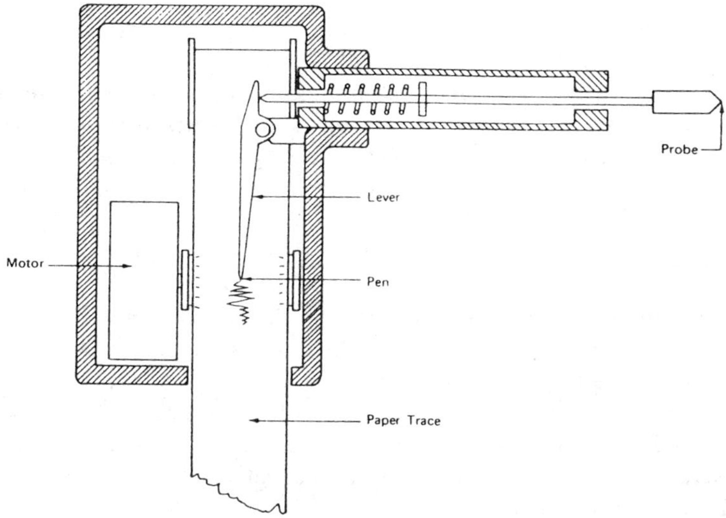

Figure 7-12 Vibrograph

This type is suitable only for low-frequency measurements with reasonably large amplitudes because of the limitations of a mechanical system. High- frequency and small amplitudes cannot be read from the trace. Because it is hand-held, some hand movement is inevitable, and this will alter the natural movement of the object. However, as a quick, simple, and cheap method of measuring displacement it is often very useful.

Strain Gauge

The strain gauge (Figure 7-13) can also be used to measure displacement. It is attached directly to the vibrating object, and the strain in the material causes a resistance change that is measured in a Wheatstone bridge. A bridge excitation voltage is supplied and the deformation produces a voltage change proportional to the strain, and hence also proportional to the displacement. The strain gauge is simply connected to a strain indicator of a type suitable for dynamic measurements, such as shown in Figure 7-13.

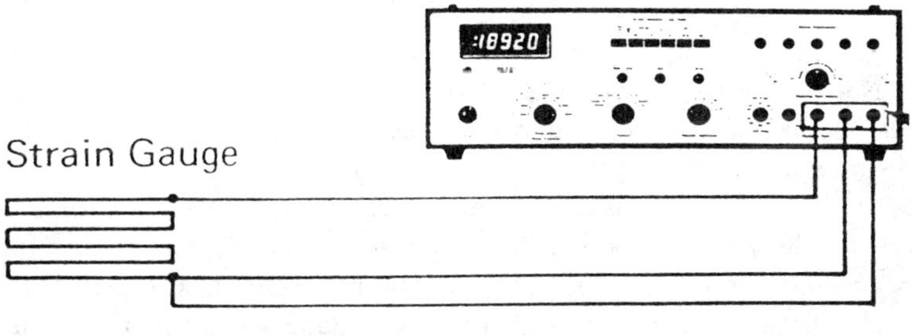

Figure 7-13 Strain gauge system for displacement measurement.

Velocity Measurement

Velocity, or the speed at which the measuring point on the vibrating object is moving, is measured with a moving element transducer, either a coil or magnet type, which touches the object, or a magnetic one, which is contact free.

Moving Element Transducer

An example of the moving element transducer is shown in Figure 7-14. When the transducer is subjected to vibration, the movement of the moving element (in this case the coil) will induce a voltage in the coil. This induced voltage is proportional to the relative velocity of the coil. This sensor is self-generating, and has a low output impedance. However, its moving parts are prone to wear, and it is fairly bulky. It is also delicate and sensitive to magnetic fields and orientation. Small velocity pickups have high internal friction and so are less sensitive. Their directional sensitivity can also change at low frequencies.

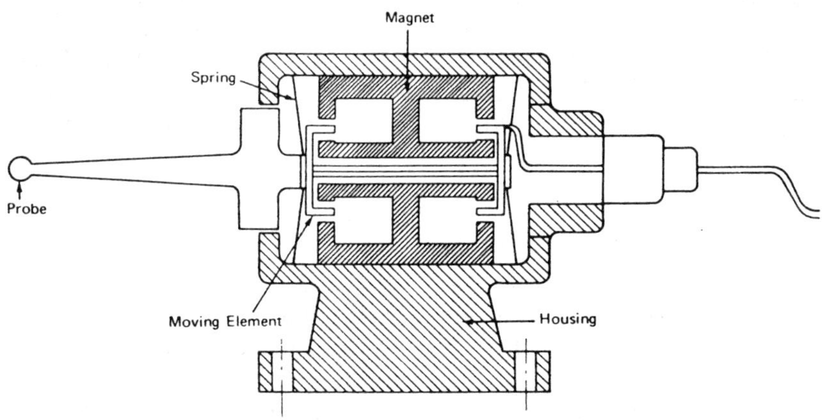

Figure 7-14 Moving element velocity sensor.

Piezoelectric Accelerometer

The piezoelectric accelerometer (Figure 7-15) has been a commonly used transducer for measuring acceleration. It is an electromechanical device that relies on the fact that deformation of a piezoelectric ceramic element by an applied force produces an output voltage proportional to that force. If a mass is attached to the ceramic element and the entire system vibrated, the force needed to accelerate the mass is applied through the ceramic element. Since force = mass × acceleration (Newton's Law), the output voltage is also proportional to the acceleration. The electric charge ¯ voltage × capacitance is also proportional to acceleration, so either voltage or charge may be measured.

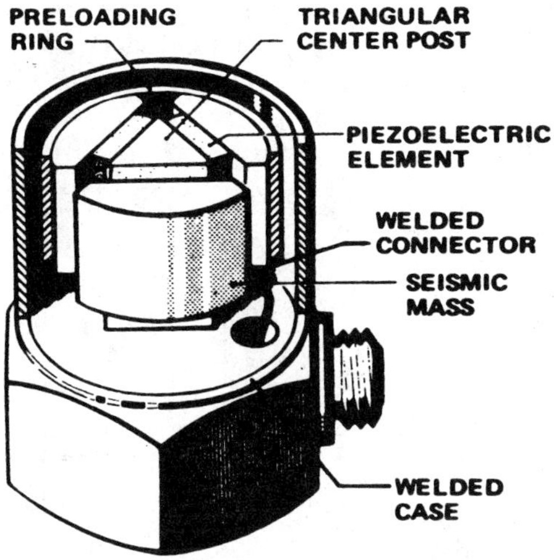

Figure 7-15 Piezoelectric accelerometer.

This type of accelerometer is self-generating with no moving parts to wear and is very rugged and compact. It has a large dynamic range and a wide frequency range, is relatively inexpensive, very reliable (having high stability), and easy to calibrate. It can be mounted in any orientation to measure the acceleration component along its axis, but it must be rigidly mounted on the test object. Care must be taken to ensure that its mass does not affect the motion of lightweight specimens. Its limitations are that it has a high impedance output and no true DC response.

The frequency response of a typical piezoelectric accelerometer will look similar to Figure 7-16. The lower limiting frequency depends on the cable, the preamplifier, and the environment. The upper limit is set by the method of attaching the accelerometer to the test object.

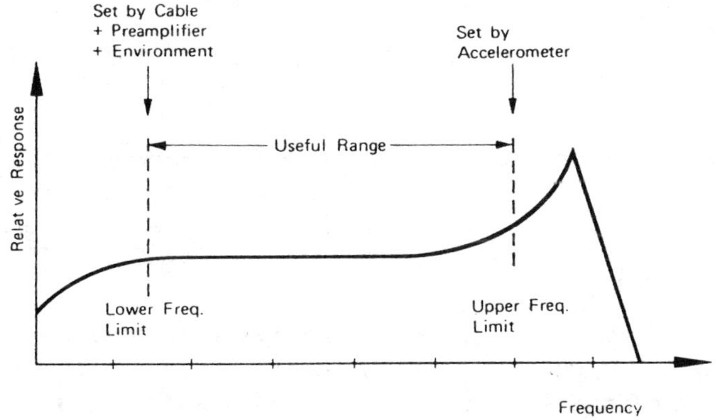

Figure 7-16 Frequency response of a piezoelectric accelerometer.

The responses with various attachment methods are shown in Figure 7-17. The best method is a steel stud, although cement, Eastman 910, gives similar results. But magnets and hand probes introduce a spring effect and lower the natural frequency of the system. Measurements are usually made up to about

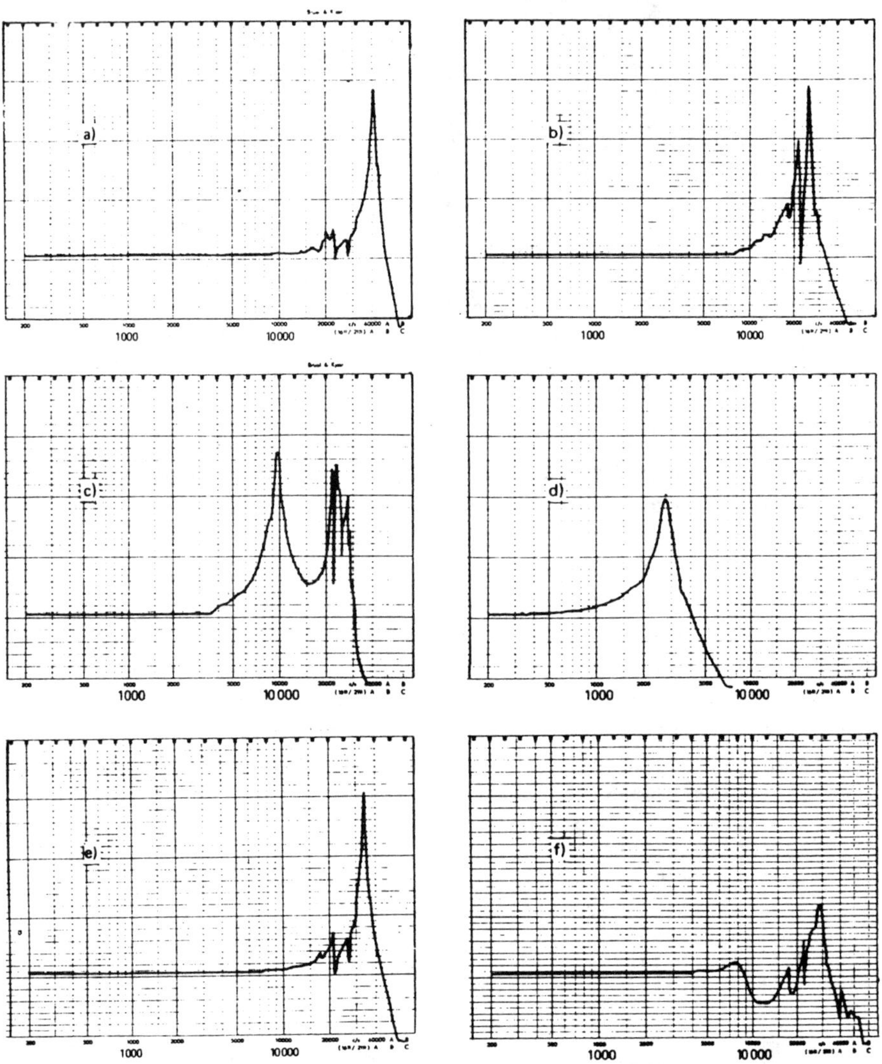

Figure 7-17 Typical frequency response curves of an accelerometer with different mounting methods: (*a*) steel stud, (*b*) isolated stud, (*c*) permanent magnet, (*d*) hand-held with probe, (*e*) was, and (*f*) soft glue (not recommended).

one third of the resonance frequency. The electrical output of a typical piezoelectric accelerometer is proportional to acceleration over the useful range (Figure 7-18). The lower limit is generally set by preamplifier noise, and the upper limit is set by the accelerometer's structural strength. The upper limit is usually many thousand g's.

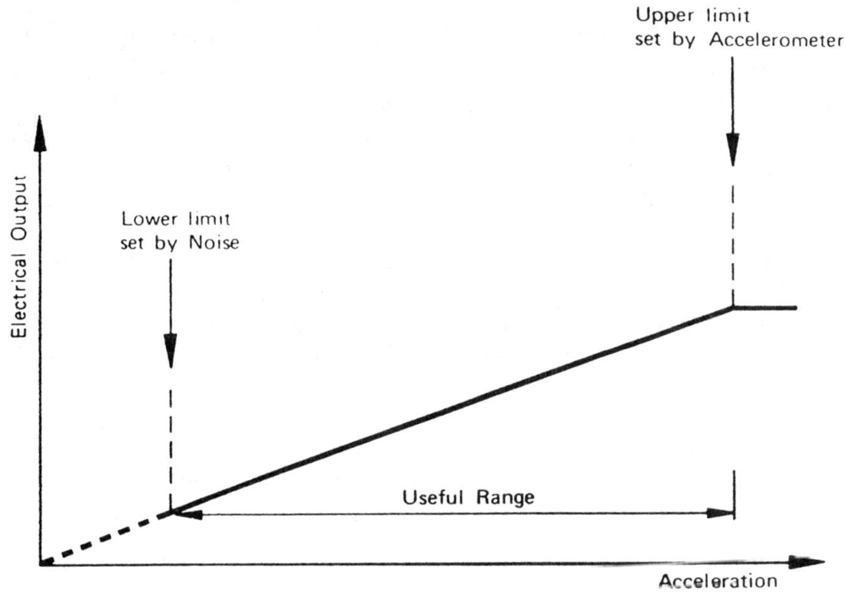

Figure 7-18 Electrical output as a function of acceleration for a piezoelectric accelerometer.

VIBRATION METERS

A system for measuring vibration consists essentially of the components shown in Figure 7-19. Considering the signal path through the instrument, the first essential component is a transducer to convert the vibration signal to an electrical

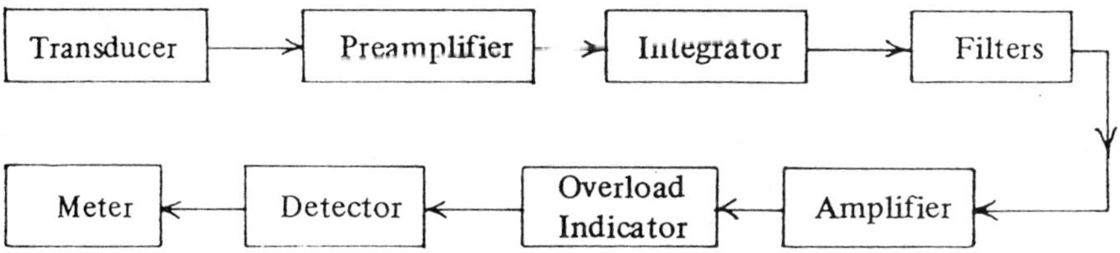

Figure 7-19 Essential components of a vibration system.

electrical voltage or charge variation. The preamplifier is required to present the proper electrical load impedance to match the transducer output, thus providing a uniform response of the electrical system across the widest possible frequency range. The integrator is not an absolutely essential component in all cases, but if velocity or displacement measurements are required from an accelerometer input, it becomes essential.

Filters are necessary to limit the frequency range of the instrument to the measurement requirement and can be provided either internally or externally to the vibration meter. The amplifier provides the dual function of matching the high electrical impedance of the filters and adjusting the signal level to meet the relatively narrow dynamic range of the detector circuitry. It is protected by an overload indicator that guarantees that the signal is not distorted from overdriving the amplifier circuits.

The detector determines which of the signal parameters—RMS, peak, or average—is displayed on the meter. Usually it will be the power content of the signal that is of interest, and therefore the RMS value of the signal is needed from the detector. Tunable filters and graphic records are available for making permanent records for future comparisons.

8 VIBRATION CONTROL

PADS AND MOUNTS

Large heavy items of motor-driven equipment such as fans, pumps, compressors, punch presses, and milling machines are prime examples of equipment that vibrates severely during operation. One method of reducing vibration transmission is to separate each machine, on mounts, so that the shock is contained. Vibration produced is blocked from traveling beyond the machine itself by isolating the equipment so the vibration will not be transmitted to the floor, structural members, or rigid sections of electrical distribution systems.

One of the simplest methods of controlling vibration and shock from machine installations is with isolation pads made of felt, cork, elastomers, steel, aluminum, or numerous combinations of these materials. Pads can be employed by simply placing them under the legs or base of the machine. In textile mills, for example, where there is excessive unidirectional vibration, pads are cemented to the floor and absorb the energy of the low-frequency vibrations that are produced. Isolating pads can also be bolted to the floor. Some pad designs require no special attachment method but are simply put down and remain in place. To prevent movement, the lower surface of the pad may be a waffle-like design that creates a suction with its contact surface. Elastomers are also vibration isolators that work in curtailing detrimental vibrations, which would otherwise be transmitted to floors, foundations, and other adjacent machines. These pads will protect a machine from both impact and vibration forces that are transmitted through the floor and foundation from other machines and from vibration caused by the machine itself. Isolators also provide a slight leveling action because of their resilience and their ability to keep the structure free from nonuniform stresses that might be produced by vibration from external sources. Most pads resist petroleum products, cleaning compounds, and deteriorating effects of sunlight and air.

Inertia blocks are also widely used when there is a possibility of shock and vibration forces being transmitted through the foundation. These blocks are concrete sections that are set in the floors or foundations and are isolated from the surrounding material by vinyl-bonded pads, which are inserted before the concrete is poured to act as an impervious barrier. Motors and machines can be mounted on massive inertia pads that are spring mounted above floor slabs. Construction of pads varies; one of several possibilities consists of a reinforced concrete deck with an outer steel frame incorporating the spring housings.

With the use of relatively lightweight motors and machines, additional isolation can be affected by short rigid sections of electrical conduit stubbed down through floor slabs, passing through an amply sized sleeve. The sleeve is then packed with fiberglass and caulked to make it airtight. The flexible conduit that extends through the lower slab continues to a remote switch gear location. Another method for isolating vibration from rigid electrical distributor systems is to stub-up motor feeder conduit adjacent to, but not in contact with, the inertia pad; then secure a feeder terminal box to the outer steel frame of the floating slab. A large hole is cut in the bottom of the box that surrounds the conduit stub, and the space between the rigid conduit and floating box is later sealed with a thick neoprene gasket.

RUBBER ISOLATORS

Because of economy, ease of installation, and replacement, rubber isolators are much preferred. Some important characteristics of rubber isolators are:

- elastic properties
- damping ability
- ability to reduce noise transmission

Rubber has an excellent damping ability, which can be expressed in the following equation:

$$\phi = \frac{\textit{vibrational energy transformed into heat}}{\textit{total energy of vibration}}$$

ϕ is the coefficient of damping, and for natural rubber ϕ will average approximately 0.25; for pressed cork ϕ will average approximately 0.12, and for concrete ϕ will average approximately 0.015. The comparison indicates that rubber will change approximately one fourth of the energy into heat and is therefore an excellent material to absorb vibration energy.

Extreme temperatures affect natural rubber characteristics, which change its isolation effectiveness. Varying temperatures between 20°F and 140°F will change the characteristics of natural rubber from 12 percent to 30 percent for each 36°F change. Extremes in temperature can have permanent detrimental effects on rubber. Rubber isolators are effective when the desirable frequency is above approximately 300 vibrations per minute. When natural frequencies below 200 vibrations per minute are encountered, ring mountings are usually recommended.

SPRING SYSTEMS

Vibration control can be accomplished by using steel springs in combination with sound-absorbing materials. Because of their effectiveness, springs are widely used in machine foundation design. Advantages of rings are:

- smoother and quieter machine operation
- reduced machine maintenance
- reduction or elimination of the effects of vibration on the structure and occupants
- a decrease in size and mass of the foundation under the equipment
- more accurate design because properties are known and readily predictable
- considered permanent for the life of the machine; properties are not affected by time
- leveling adjustment is possible in spring units
- versatility of installation

Disadvantages of springs are:

- higher initial cost of construction in some installations
- requirements for maintenance of the spring units
- noise transmission and no damping ability

Air Springs for Vibration Control

The ultimate design goal is for vibrationless equipment, but in most cases this is unattainable. Typical problems involve isolating areas that cause undesirable forces. One of the elements available for solving this problem is the air spring. Industrial air suspension systems are used to:

- isolate structures from vibrating masses
- control the frequency of vibrating masses
- keep a suspended platform at a given height under variable-load conditions

The air spring is basically a metal container with a rubber liner filled with air. It is a stored energy medium that is designed to support a given load. The important characteristics of any spring in a suspension system are:

1. Transmissibility—the ratio of force transmitted through the mounting to the disturbance force. Air springs have low transmissibility coefficients.
2. Spring rate—the change in load per unit of deflection. The spring rate is directly proportional to the load on the spring and to the pressure of the air in the spring assembly. This is the result of a constant natural frequency suspension, regardless of the load.
3. Load carrying ability—the load range between maximum load and pressure and minimum load and pressure, for which the ratio is 5:1. Air springs have an inherent feature because their load carrying abilities and ratio can be varied by varying their air pressure.

Current designs in air springs are bellows, rolling lobe, bellobe (a combination of bellows and rolling lobes), restrained rolling lobe, filling supported sleeve, and hydropneumatic.

Air-suspension systems range from simple to complex depending on the application. If small variations in height are not important and the pressure is constant, then a tire valve can be installed to vary air pressure. The air pressure should be checked periodically and adjusted if necessary. If the same pressure is required in all air springs of a particular system, a pressure regulator valve can be installed to keep the system at constant height. If variable height is desired, automatic leveling valves can be installed. Occasionally it may be necessary to deflate air springs simultaneously so the entire controlled mass rests on solid supports. For this result, a series of check valves may be installed; when the main valve is opened, all the air springs deflate.

An additional advantage of air-spring systems is that time has no harmful effect on them. Air columns do not deteriorate with age, nor do they tend to set as mechanical springs might; however, costs of these systems are somewhat higher. Pneumatic systems are easy to install because there is no need for special foundations. Devices are set in place under a machine's support points and connected to the plant air supply of 85 to 100 psi. Readily available units can handle equipment weights from a few hundred pounds to several hundred tons.

ADJUSTABLE DAMPERS

Adjustable dampers operate on the principal that the workpiece can be considered as a part of an entire machine structure and therefore can be the source of chatter. The side opposite the cutting tool contacts the damper with the weakest portion of the workpiece. This absorbs vibratory energy and increases the dynamic rigidity of the machine structure.

The internal resistance of the machine structure depends on the machine structure's dynamic rigidity. Experience has shown that the damping ratio in machine tools is not sufficient to resist resonant vibrations.

When cutting down on speed or the feed rate, productivity has to be sacrificed. With an adjustable damper this problem is eliminated. Other advantages of the external adjustable damper are the cutting process dynamics remain unchanged and the required chatter control is obtained by changing the machine structure.

In this type unit a restraint developed by a piston arm and spring force allows the mass to move up and down. An adjusting screw located on top of the system is used to adjust the amount of spring desired. When the roller is in rotation and the cutting edge passes the roller, the roller is automatically lifted to clear the tool and cut and is then repositioned to place the roller on the new surface. Limit switches, which are located in the lower part of the damping system, control these motions. When these rollers are not in contact with the workpiece, the preload on the spring can be easily retained by a control switch that is located on the piston arm. In the hydraulic type system, there is a flow control valve that restricts the repositioning speed and in turn avoids shock vibrations when the rollers come in contact with the workpiece.

For certain conditions, several sets of springs may be needed to meet various conditions. By using several sets of springs the phase angle, which depends on the damping constant and spring rate, can be carefully controlled. The spring is preloaded to suppress disturbances during the period following the cutting tool's contact with the workpiece. During this period the system approaches a steady-state condition.

Coulomb damping is friction between two sliding pads and is often used because it is easy to apply and economical. This type of damping has a quick response. Lubrication is provided between the sliding parts and, as a result, linear viscous damping occurs. There are three screws on each damper to adjust the degree of damping required. In order to prevent any kind of disturbance, friction between the sliding parts should be kept to a minimum; hence the use of lubrication. However, friction can never be entirely eliminated, and a small amount of it is necessary for a system to reach steady state. The dominant factor in the system is stationary; the sliding parts contribute nothing to the system.

The dampers can be located anywhere on a workpiece, but for a lathe they are generally hung on the frame parallel to the center line of the spindle. Usually on a shorter piece of work only one damper is used, but more can be employed if needed. If several dampers are used on a roll and each one is manually adjusted, a skilled technician should make the adjustments. To avoid this, sensor devices are attached to the dampers and all of the dampers can be adjusted from a center controller. It is important to check the condition of the lubricant; stick-friction should not be used in the system in order to avoid any serious disturbances. The damper can be used in many different applications without any difficulties.

TYPES OF VIBRATION MOUNTS

Cork has been used in many fields of industry as a vibration isolation material. A popular type of cork is made of pure granules compressed together and baked under pressure to achieve a controlled density. Cork materials are used mainly under concrete foundations as illustrated in Figure 8-1. The cork will remain reasonably durable under exposures to acids, oils, and temperatures between

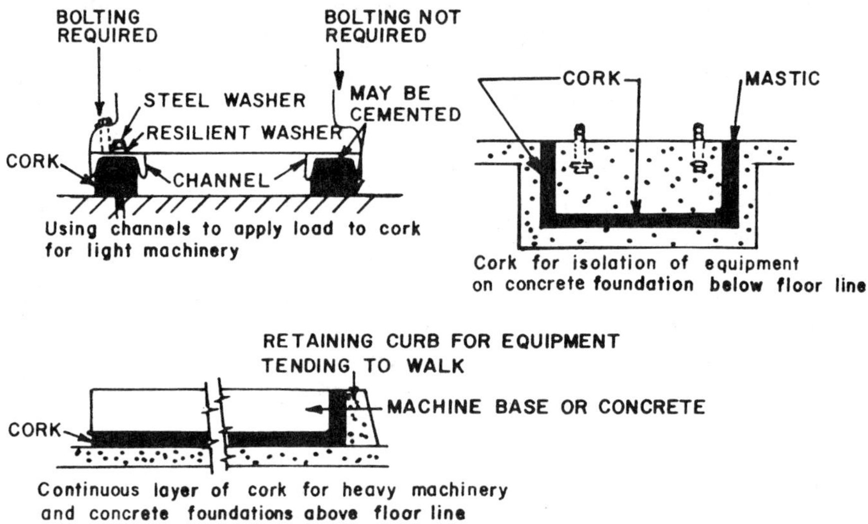

Figure 8-1 Cork applied under concrete foundations.

between 0°F to 200°F; however, it will be affected upon contact with strong alkaline solutions. Cork will rot, however, from repeated wettings and dryings. As a vibration isolator, cork is limited to frequencies above 1,800 cycles per minute. Because of a great degree of damping in cork, the natural frequency cannot be obtained from the static deflection. As an alternative, the natural frequency can be obtained through tests by vibrating the cork under various loads to find the resonance frequency.

Rubber has also been mentioned as a vibration isolator and is useful for frequencies above 1,200 cycles per minute. Alkali solutions or acids will not affect rubber, but degradation problems could arise if it is exposed to sunlight. For natural rubber, the temperature range is from 50°F to 150°F; for neoprene from 0°F to 200°F. For applications that expose the rubber to oil, neoprene is more suitable. As rubber ages, it gradually loses its resiliency. The useful life span of a rubber mount is approximately five years under impact applications and seven years under nonimpact applications, although it will retain its sound-insulating properties for much longer. Individual molded rubber mounts are economical only with medium- and lightweight machines because heavier capacity mounts approach the cost of the more efficient steel spring isolators.

isolators, illustrated in Figure 8-2, provide deflections up to 1.25 inches. Rubber and other isolators provide maximum deflections on the order of 0.25 inch. Some steel spring isolators can even reach deflections of 10 inches. In many other materials costly trial and error tests are usually required, but this is not necessary with steel springs because they closely follow the equations of vibration control.

Steel spring isolators are usually equipped with adjustable snubbers because steel springs contain no damping. Damping is basically helpful in restraining the movement of resiliently mounted machinery but it does reduce the isolation qualities of the mount. Many of the steel spring isolators used by industry have built-in leveling bolts that eliminate the use of shims upon installation. Often, rubber sound-isolation pads are used in conjunction with steel springs because high-frequency noises have a tendency to by-pass steel springs. Table 8-1 illustrates the practical range for various isolation materials at different equipment speeds.

Table 8-1 Relative Effectiveness of Steel Springs, Rubber, and Cork in the Various Speed Ranges

Range	rpm	Springs	Rubber	Cork
Low	Up to 1200	Required	Not recommended except for shock[a]	Unsuitable except for shock[a]
Medium	1200-1800	Excellent	Fair	Not recommended
High	Over 1800	Excellent for critical jobs	Good	Fair to good

[a]For noncritical installations only; otherwise, springs are recommended.

BASIC INSTALLATION PRINCIPLES

When mountings are installed to cut down machinery vibration, the following principles should be observed to ensure maximum efficiency:

1. All parts of the unit of machinery to which the mountings will be applied must be placed on a common rigid base.
2. The entire unit of machinery should be insulated with a suitable mounting.
3. To ensure full effective insulation, elastic connections should be made to the unit as shown in Figure 8-3.
4. If required, grounding should be installed to conduct away any static electricity.
5. If the unit of machinery is tall or unstable, it should be fitted with an enlarged base with a thick, rigid plate.
6. Coupling for elastic force transmission should be done.

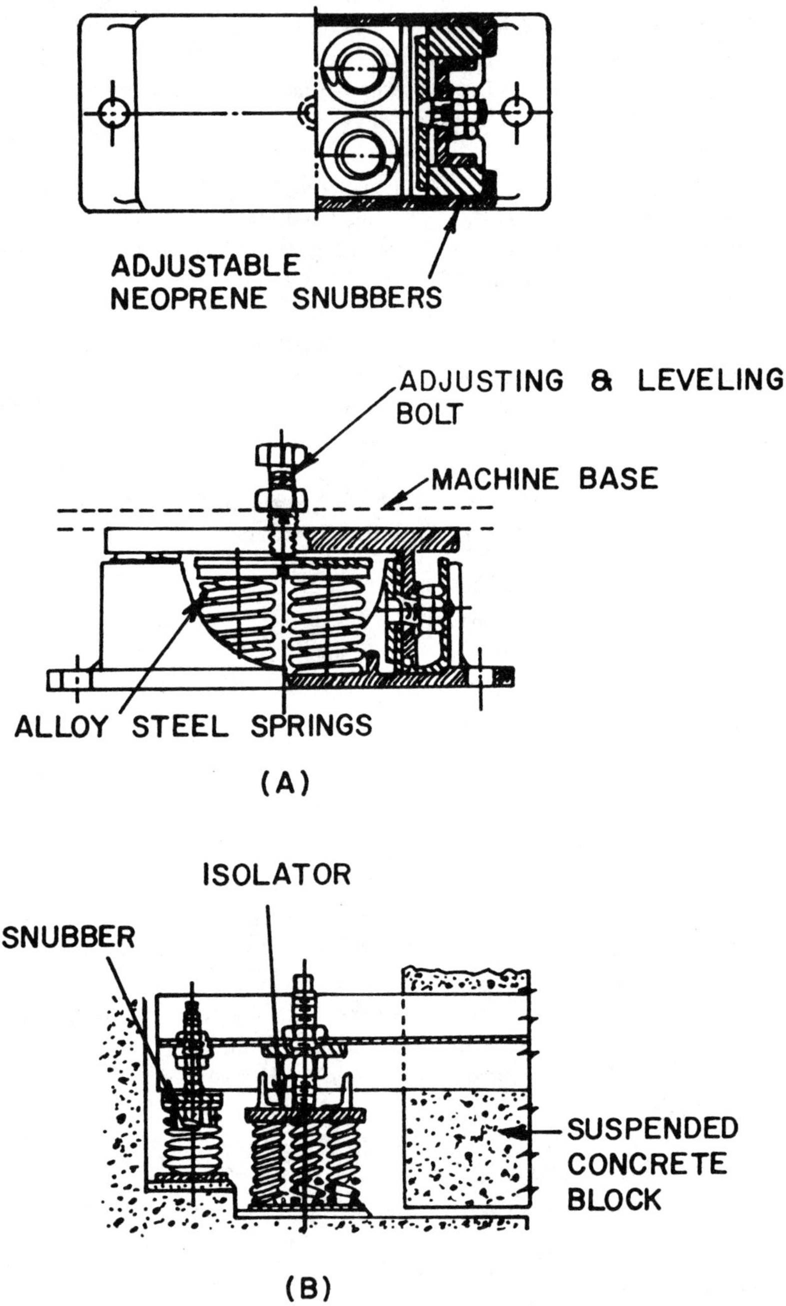

Figure 8-2 Steel spring isolators.

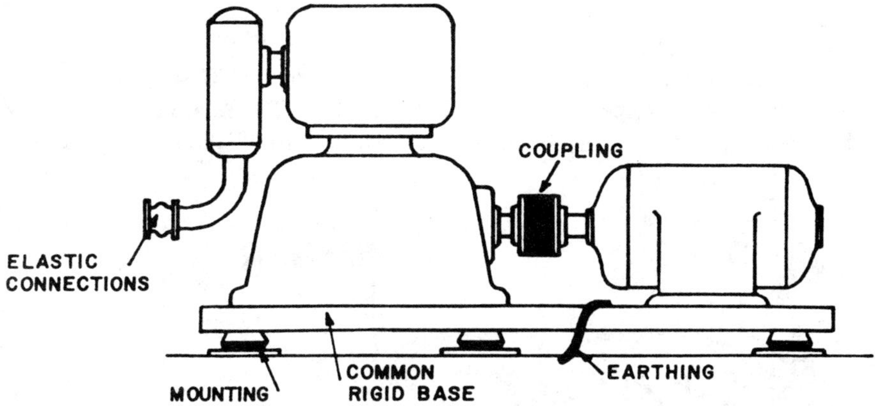

Figure 8-3 Elastic connections are made to ensure full insulation.

SEVERAL TYPES OF INDUSTRIAL MOUNTINGS

Type 1

This mounting, illustrated in Figure 8-4, consists of a cylindrical rubber body with steel plates tightly vulcanized to each end. Each plate has a threaded center hole for attachment. Upon installation, the top steel plate is bolted to the machine

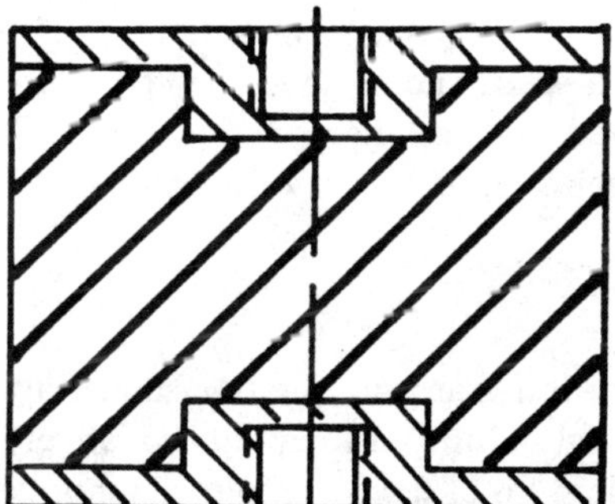

Figure 8-4 Cylindrical steel and rubber type 1 mounting.

machine base and the bottom steel plate is anchored to the floor. This type of mounting is specifically suitable for vertically loaded units of machinery that only make small sideways motions when in operation. The mounting is flexible and can be made even more so by tilting the insulator; therefore, insulation is achieved through shear and compression rather than just compression. This mounting can be applied to such machinery as converters, fans, pumps, and electrical machinery.

Type 2

This mounting, shown in Figure 8-5, is made of two U-shaped steel components with rubber tightly vulcanized between them. There are two holes under the inner shank to improve resilient qualities. To install, the top U-plate is

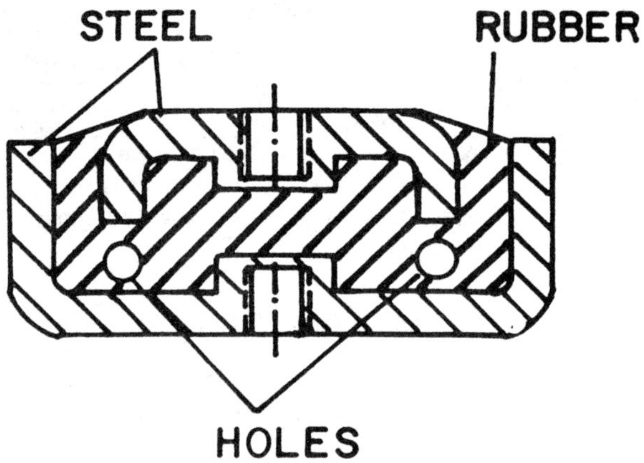

Figure 8-5 Type 2 mounting.

is bolted to the base plate of the unit of machinery and the bottom U-plate is anchored to the floor with an expansion-shell bolt. This type of mounting is especially good for high-speed and heavy machinery because it provides a stable base. It can be applied to machinery such as punches, presses, carpentry machines, weaving machines, and transformers.

Type 3

Figure 8-6 illustrates an antivibration mounting designed to give high resilience and good lateral stability. Mounting is relatively easy since installation height is low. On the bottom of the mounting is a thin layer of rubber that will prevent the machinery from moving across the floor. The friction supplied between the mounting and the floor is usually sufficient to make bolting unnecessary. The top metal cover is designed to protect against oil spills. This type of mount is made for both light and heavy low-speed machines, which are generally difficult to insulate, and can be applied to converters, pumps, fans, combustion engines, rolling mills, presses and punches, paper machines, and various domestic machines.

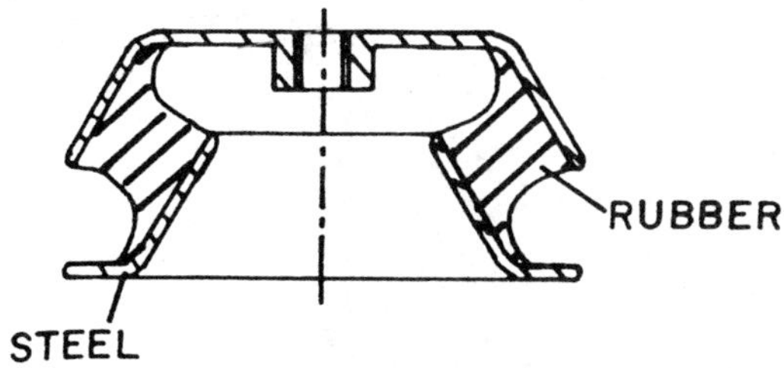

Figure 8-6 Type 3 mounting.

Type 4

Another antivibration mounting, illustrated in Figure 8-7, consists of two steel profiles of unequal size with blocks of rubber tightly vulcanized between them. When the rubber undergoes shear stress, the mounting provides

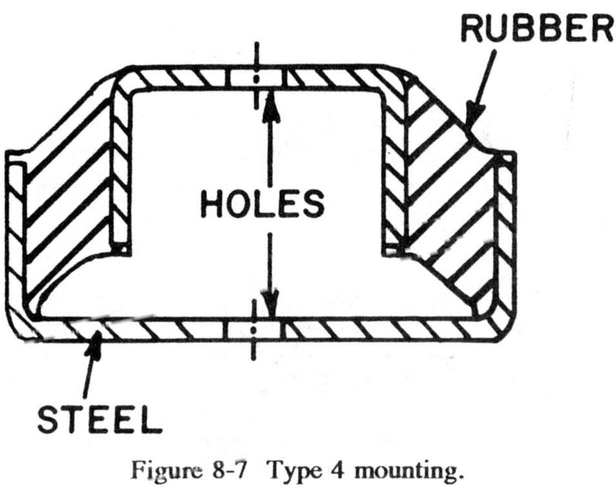

Figure 8-7 Type 4 mounting.

resilience. This insulator is used when resilient, soft mounting is needed for low loads. It is applicable to low-speed machines and light fans as well as various instruments.

Type 5

Another type of mounting is constructed of two identical steel profiles with rubber snugly vulcanized between them (shown in Figure 8-8). It provides

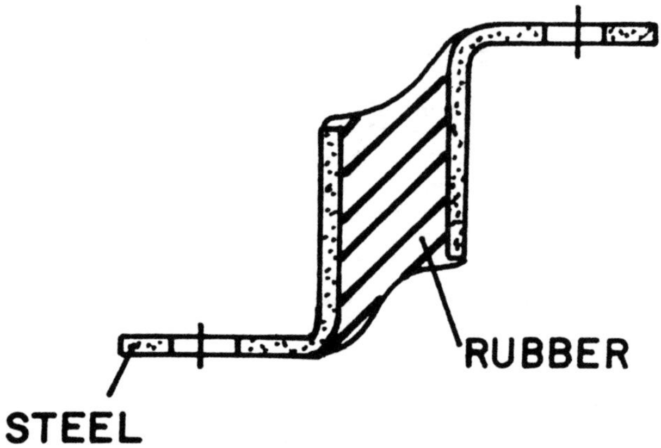

Figure 8-8 Type 5 mounting.

simple resilience when the rubber undergoes shear stress. This insulator is applicable when a soft, resilient mounting is needed for low speeds.

Type 6

This type of mounting, illustrated in Figure 8-9, consists of a cylindrical rubber component vulcanized to two steel intermediate plates. Installation of the mounting is not complicated and it can absorb large vertical forces without much deformation. It will also give good stability since its installation height is low.

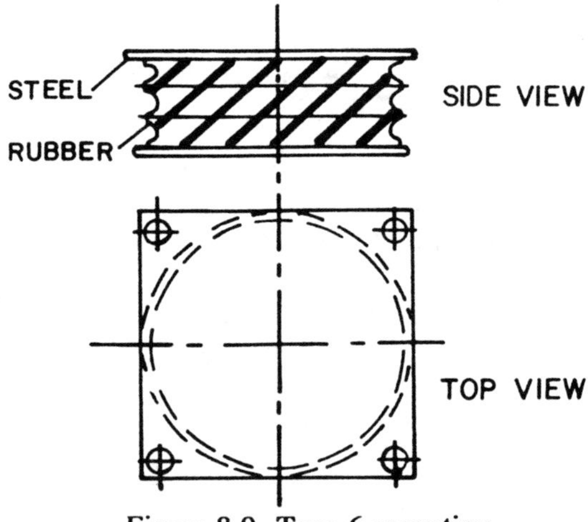

Figure 8-9 Type 6 mounting.

Providing insulation against horizontal vibrations, it has a high deflection capacity in the horizontal direction, while deflection is relatively low in the vertical direction. In light of these particular deflection properties, we can see that low interference speeds can be horizontally insulated while obtaining a subcritical mounting in the vertical direction with a safe distance to the resonance point. By connecting two parts in series, large angular movements can be absorbed and a larger vertical resilience can be obtained. This type of mounting is used for such applications as heavy machinery (crushers) and screening plants.

Type 7

The mounting shown in Figure 8-10 provides stability in all directions and is equipped with a leveling screw for varying the height of the unit to the floor. It is installed by first fixing it to the machine and then balancing the machine

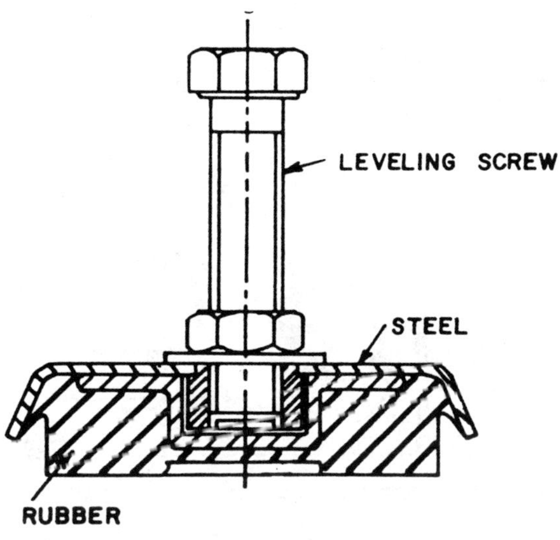

Figure 8-10 Type 7 mounting.

machine horizontally by using the leveling screw. Finally, the level position is anchored with the locking nut. There is a rubber sole at the bottom of the mounting, which provides adequate friction so the machine will not move across the floor.

This type of mounting, which has good lateral stability, is particularly useful for insulating high-frequency vibrations. It is especially applicable to lathes, punches, milling machines, grinding machines, printing machinery, and presses.

Type 8

A bumper-type vibration insulator, illustrated in Figure 8-11, consists of a cylindrical rubber component that is snugly vulcanized to a square steel

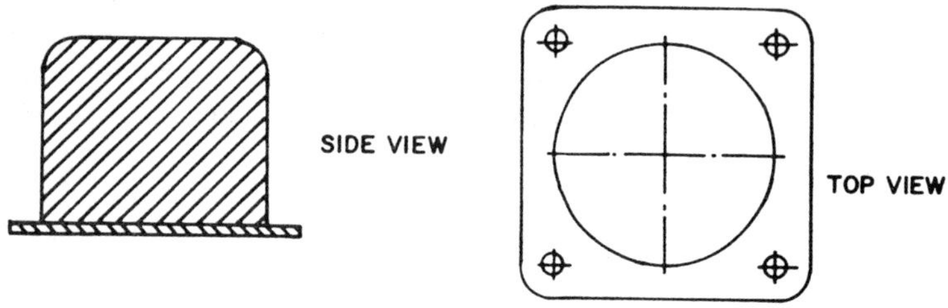

Figure 8-11 Type 8 bumper isolator.

plate. There is a hole at each corner of the metal base plate for attachment. The rubber component absorbs as much energy as possible and reduces transmitted impact forces. This isolation ensures efficient damping for moving machinery and machine parts. Considerable energy can be absorbed by the resilient rubber bumper. Two bumpers can also be used on a machine to give added resilience (as shown in Figure 8-12).

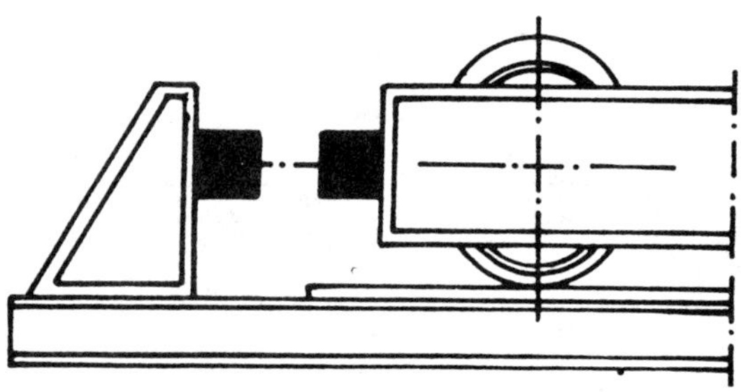

Figure 8-12 Application of the bumper-type isolator.

Two concentric cylindrical metal sleeves with rubber vulcanized between them are designed to take up torsional movements and radial and axial loads. These sleeves can be used to reduce bearing problems in vibrating constructions because the rubber takes the complete movements, so bearing maintenance is not needed. Sleeves that have very good vibration and sound insulating capacity can be applied to machinery such as linkage systems in cars, tractors, railway wagons, bulldozers, agricultural machinery, and vibrating screens. A proposed mounting of sleeves is illustrated in Figures 8-13 and 8-14.

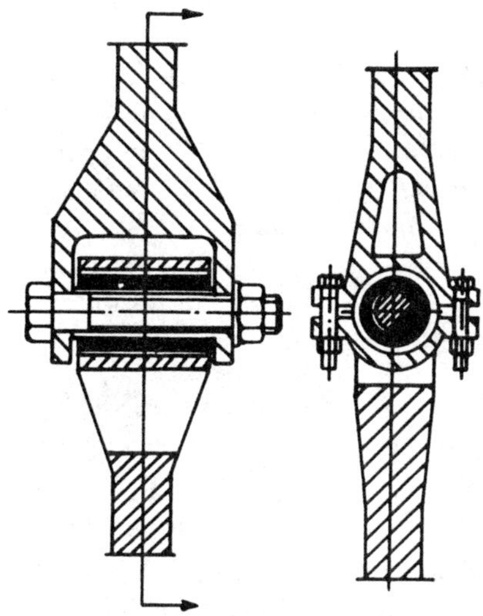

Figure 8-13 A proposed mounting of sleeves.

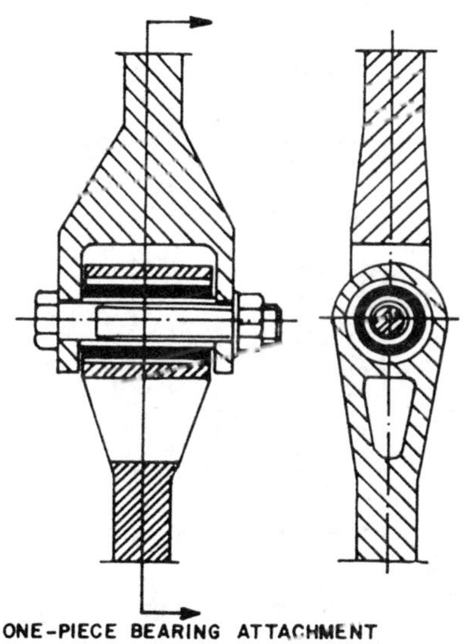

Figure 8-14 A proposed mounting of sleeves.

SHAFT COUPLINGS

The schematic of a vibration-insulating shaft coupling can be seen in Figure 8-15. It is made up of a U-shaped rubber ring that is tightly vulcanized to two steel rings on either side of it. When in operation, the rubber U-shaped ring will transmit torque from one of the steel rings to the other, thereby delivering the power without vibration.

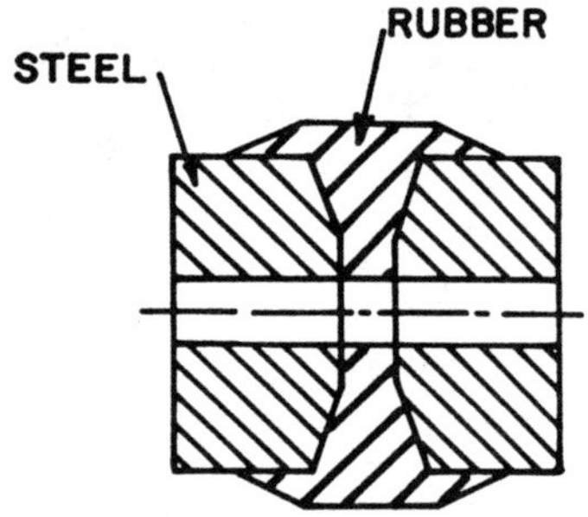

Figure 8-15 A vibration-insulating shaft coupling.

Another type of shaft coupling is shown in Figure 8-16. This coupling has special metal stops that absorb any uncommon torsional stresses. It also has shaft flanges securely screwed to the rubber coupling.

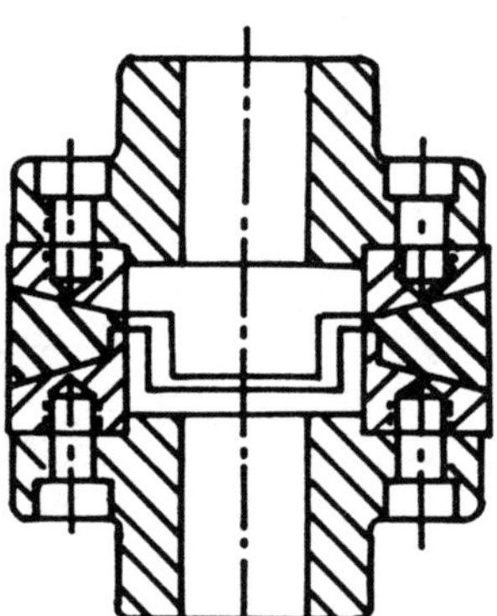

Figure 8-16 Another type of shaft coupling.

In general, shaft couplings protect machinery by providing for smooth, silent, and well-balanced running. They can level out any small deviations between the shafts, which could prove valuable in cases where shaft alignment is very difficult. The type illustrated in Figure 8-15 is generally more applicable for smaller machines, and the type illustrated in Figure 8-16 is more suitable for heavier machinery, such as paper machines and excavators, in that it allows for greater power transmission.

VIBRATION PLATES

Two popular versions of this unit are the single and the double plate. The single-plate version has uniform ribs on one of its sides; the double-plate version has ribs on both of its sides, which are at angles to each other. These plates, illustrated in Figure 8-17, are made of oil resistant rubber.

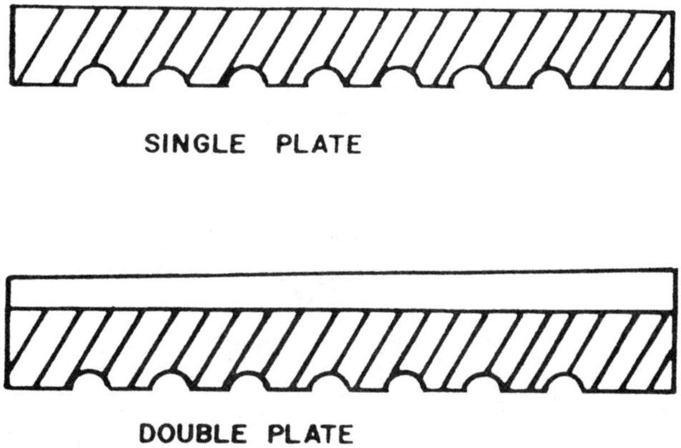

Figure 8-17 A typical vibration plate.

These plates are used in industry for relatively simple installation requirements. Best suited for damping a high-frequency vibration, they can be arranged in series to give added cushioning. If mounting is to be permanent, a direct contact between the base and the machinery must be avoided. A mounting example is shown in Figure 8-18.

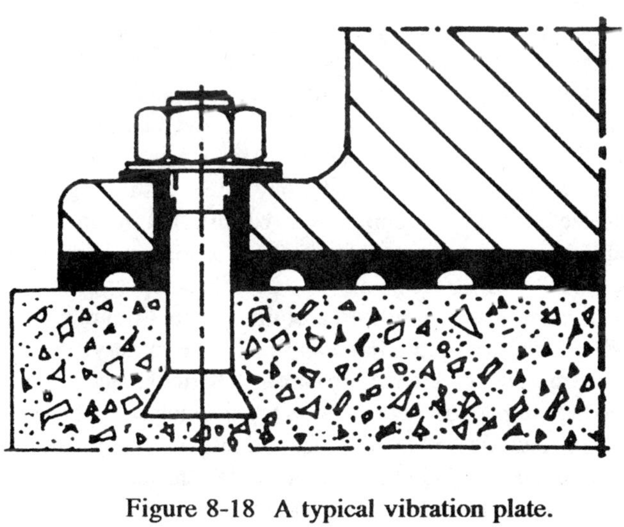

Figure 8-18 A typical vibration plate.

BLOCKS

This antivibration component is just a simple rubber block with recesses in it for installation (see Figure 8-19). Blocks are easy to fit because they are placed loosely between the machine and the base. The friction produced between

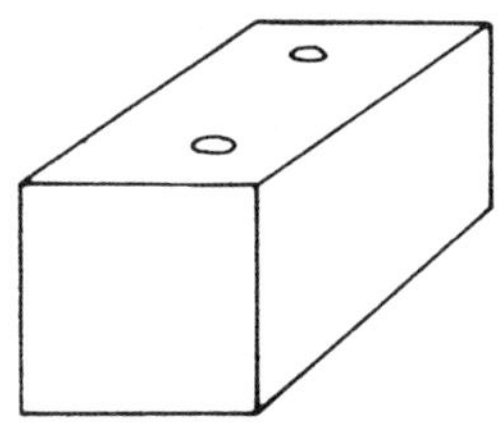

Figure 8-19 A typical antivibration block.

the rubber and the base is sufficient enough to keep the machine from moving. There are four recesses in the block shown, which can be used as sockets for pins or dowels to be positioned to the machine and base.

Blocks can be used to insulate heavier machines with low interfering frequencies. Particularly useful for large composite machines that are on a common girder frame, they are also applicable to many types of machinery including converters, mixers, gear wheels, and rolling mills.

Some Applications for Isolators

Properly designed mountings allow the installation of some of the heaviest machinery in penthouses, near offices, and many other areas where vibration and noise would be a nuisance. The Federal Housing Administration (FHA) has approved the use of heavy machinery on many penthouse installations provided it is insulated with steel spring isolators. When heavy machinery is installed on the upper floors of such complexes, careful consideration must be taken to prevent the transmission of vibration to several floors below when a ceiling, wall, or fixture has the same natural frequency as the vibration. In cases of such resonance vibration, the result can be a disturbing noise. When mountings are selected, heavy machinery can be installed in more economically constructed and lighter structures with fewer noise problems. Machinery can also be installed in older buildings that may not have been specifically designed to handle such equipment with the aid of carefully selected mounts. Figures 8-20 and 8-21 show the results of mountings applied to machinery.

In such structures as anechoic room installations, steel spring isolators resting on rubber sound pads are used to stop the transmission of structure-borne vibration and the resultant noise to these rooms. In many air-conditioning installations, the noise and vibration that travel through the piping poses annoying problems. However, if the refrigerating compressors are installed on

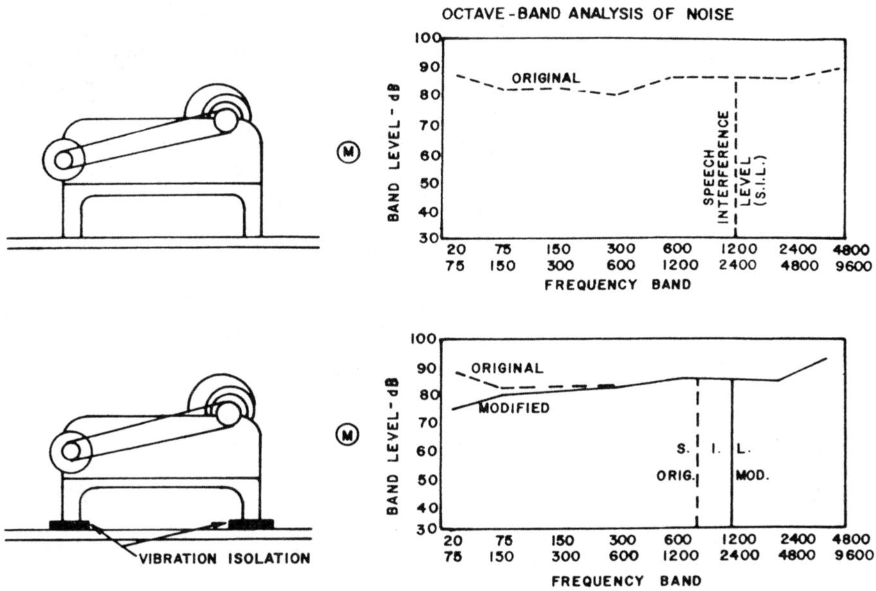

Figure 8-20 Results of mountings applied to machinery installed in buildings that were not constructed to accommodate them.

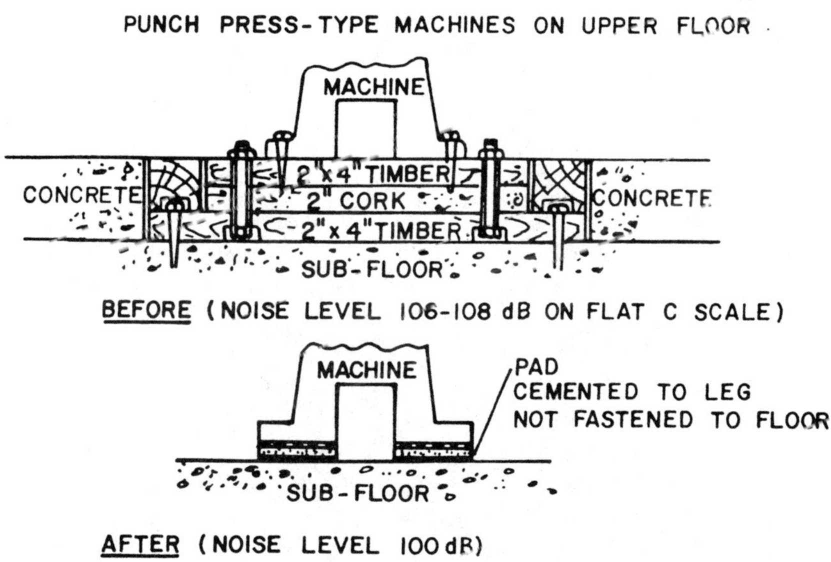

Figure 8-21 Results of mountings applied to machiner installed in buildings that were not constructed to accommodate them.

mountings, allowance should be made for flexibility in the intake and discharge piping to reduce the transmission of vibration. Providing flexibility in the piping itself can be achieved by running the piping a distance equal to 15 times the pipe diameter both horizontally and vertically, and then attaching the piping to the structure. Flexible metallic hose can also be used.

When the piping is suspended from isolation hangers it is given added protection. On the discharge and intake sides of water pumps, flexible rubber hose (approximately three diameters long) should be used. On intake and discharge fans, flexible ductwork should always be used.

9 VENTILATING SYSTEMS

Air-conditioning systems vary greatly in design and purpose. For example, for heating a building, a heat pump must be included in the design. When humidity control is desired, reheating coils following an evaporator cooling system are included. In industrial sites or hospitals requiring air sanitation, electric precipitators, wet collectors, or ultrasonic agglomerators may be necessary. These designs may also require air sterilizers and activated carbon absorbers for odor removal. When only ventilation is necessary, a forced-air draft system is employed whereby a fan directs filtered air to designated locations in the building.

Any of the many components of an air-conditioning system may be a cause of unwanted noise generation or vibrations. Adequate noise control of these systems can be achieved by careful examination of the possible causes and by providing proper isolation and insulation to the primary units in the system.

Ventilating Systems

In general, ventilating systems consist of a motor-driven blower that directs air through a plenum into headers for distribution to various parts of a building. Often the fan is located in the plenum to isolate noise. Air flow is regulated by the difference in pressure between the two terminals of the duct network. Designs are classified according to low-, medium-, and high-pressure systems. The maximum allowable static pressure at any supply or return opening for each type is given in Table 9-1.

Two principal types of noise are normally encountered in simple ventilating units—fan noise and air-flow noise. Fan noise can be subdivided into two components: rotational and vortex. The rotational component is associated with the impulse transmitted to the air each time a blade passes a fixed position. Hence, it is a series of discrete tones at the fundamental blade-passing frequency and is a function of the harmonics. The vortex component of noise is attributed largely to vortices shedding from the fan blades. This occurs due to air turbulence

Table 9-1 Classification of Ventilating Systems

Type System	Maximum Static Pressure (inches of water)
Low pressure	0.5
Medium pressure	3.5
High pressure	6.0

turbulence caused by the wind stream passing over system components such as elbows or filters. Medium- and high-pressure systems are often plagued with turbulency noise generation.

For short ducts, noise generation is derived primarily from the fan. However, for long distribution systems, in addition to air turbulence, other sources may exist. For example, fan noise may become masked by sounds of air turbulence and radiations from entrance ports. Wall and panel vibrations originating from the operating machinery may also exist and might result from equipment failure such as fan imbalance or bearings. In addition, brush or magnetic noise may develop in the fan.

Air-Conditioning Systems

The location of possible noise trouble spots becomes complicated with more elaborate systems because there are a large number of components that could be the cause. The major units that generally require attention are compressors, condensers, cooling towers, evaporators, and pumps and piping.

Compressors

Compressors must be located relatively far from the distribution system. They are often mounted on an elaborate mounting block on the top floor of a building. To provide noise control, a floating concrete floor is constructed with a suspended ceiling with acoustical paneling directly beneath it.

Condensers

The two types of heat exchangers, free and forced convection, are noise radiators that cannot be readily isolated in a special housing, as maximum surface exposure to air is necessary for high heat-transfer efficiency.

Cooling Towers

These units, generally having several hundred cubic feet of volume, are used for cooling water that has been used in heat exchangers to remove heat from the compressed refrigerant. The heat is removed by partial evaporation into the atmosphere. The process is enhanced by tube-axial or centrifugal fans, belt driven by hermetically sealed motors.

These can be extremely noisy units with severe vibration problems. They are generally mounted on the building's roof with a parapet often placed around them to reduce airborne noise. Vibration pads must be included as part of the mounting design.

Evaporators

Noise is produced by these units because the two-phase, liquid-gas mixture exiting the capillary causes a high-pitched whistle.

Pumps and Piping

Piping from pumps can undergo severe vibrations or cavitation, particularly when not properly suspended.

NOISE CRITERIA

Federal agencies and organizations have established criteria for acceptable noise levels. Different parameters and values have been used by them. Unfortunately, many different rating procedures have evolved for estimating noise levels.

Loudness and Loudness Level

Loudness level is the logarithm of the quantity of loudness. Both are a means of rating noise based on a person's judgment. The procedure for calculating loudness rates and relative loudness is standardized both in the United States and internationally.

Speech Interference Level

A calculation procedure that rates steady noise according to its interference with communication is referred to as the speech interference level.

NC and PNC Ratings

Noise criteria (NC) curves are used extensively in rating noises in buildings and in the rating of air-conditioning noises in particular. Preferred noise criteria (PNC) curves are essentially a modified NC plot.

Figure 9-1 shows the NC curves. NC values apply to steady noises and specify the maximum noise levels permitted in each octave band for a specified NC curve. NC curves have been used cautiously in the past because they tend to underestimate actual sound-pressure levels. This is particularly true for ventilating systems in which fan or duct turbulence is a major problem. To design for an acceptable noise quality, it is common to lower levels 5 dB in both the very low- and high-frequency bands.

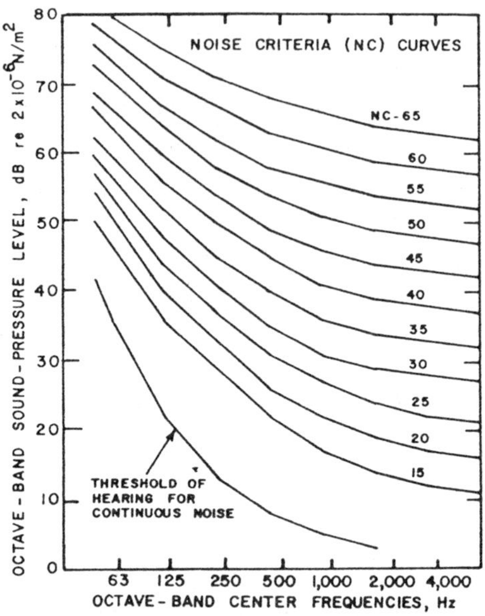

Figure 9-1 Noise criteria curves.

PNC curves (Figure 9-2) have values approximately 1 dB below the NC values in four-octave bands, to compensate for the NC inaccuracy (at 125, 250, 500, and 1,000 Hz for the same curve ratings). In the highest three bands and the 63 Hz band, values are 4 to 5 dB lower.

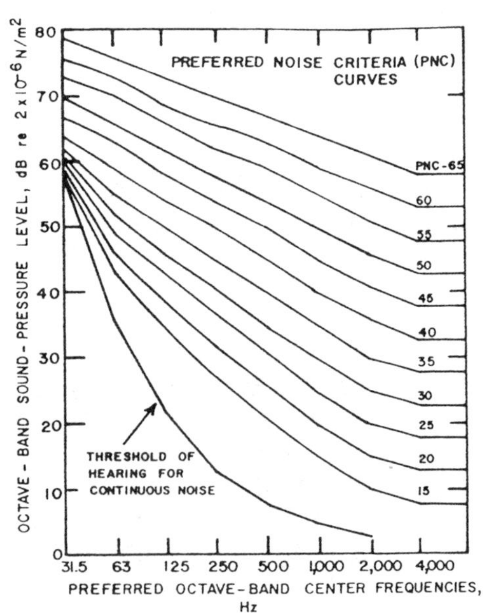

Figure 9-2 Noise criteria curves.

Noise Pollution Level (NPL)

NPL is a procedure for evaluating noises from several sources and fluctuating or intermittent sounds. The noise pollution level (L_{np}) utilizes data (either A-weighted noise levels or loudness level) that are recorded over a sufficient length of time to establish the statistical nature of the total noise exposure. The noise pollution level is expressed as:

$$L_{eq} = L_{eq} = 2.5\ \sigma\ dB\ (NP)$$

where: L_{eq} = mean-square, A-weighted sound level over some sample time, dBA

σ = standard deviation for the A-weighted sound level data

A-Weighted Sound Level (L_A)

This sound level is a means of correlating speech interference level and NC or PNC levels. This parameter is directly measured by a sound level meter that has an electronically modified frequency response referred to as A-weighting. The units are in decibels, and generally an A is denoted to signify the weighting scale (that is, dBA).

Insertion Loss

This term is used to denote the noise-level reduction as measured at a specified point in a channel after a duct silencer is positioned ahead of the measuring station.

Transmission Loss (L_{TL})

Referring to the acoustical signal reduction of a duct silencer, transmission loss is measured through a standard test by positioning a loudspeaker at the entrance of the duct network and a microphone at an exit station. Tests are made with no air flowing through the system.

Dynamic Transmission Loss

Referring to the signal reduction of a duct silencer as air passes through the channel, this parameter is generally much less than transmission loss because of air turbulence.

Different methods have been adopted to account for the effect of noise in a working environment, with recommended exposure levels varying. Standardization is not yet attained in noise evaluation. Criteria depend not only on the method of evaluation but on the specific environment; therefore, a range of ratings is usually indicated. For areas requiring a high-quality environment, the

the lower ranges are recommended. Where economy or physical limitations exist, the upper edge should be employed.

When employing PNC or NC curves, the sound-pressure level in all octave frequency bands should not exceed levels denoted by the appropriate curves. In practice, the noise level in one octave band is permitted to exceed the corresponding value on the specified criterion curve by no more than 2 dB.

The U.S. Department of Housing and Urban Development has established guidelines for noise exposure levels. The interior noise exposure in new and rehabilitated residential construction should not exceed 45 dBA for more than an accumulation of 8 hours in any 24-hour period. Guidelines for speech interference levels are also outlined. Figure 9-3 illustrates these recommendations.

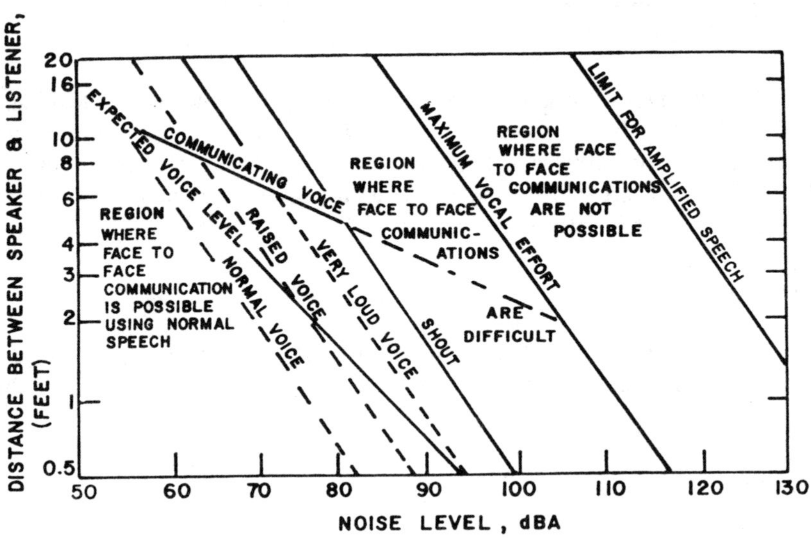

Figure 9-3 Graphic guide to allowable noise exposure levels.

MEASUREMENT TECHNIQUES

The sound level meter consisting of a microphone, amplifier, calibrated attenuator, a series of frequency-response shaping networks, and an indicating meter is most often used to measure noise levels. Weighting scales in American instruments are usually denoted A, B, C, and Flat. The C network is used to limit the low- and high-frequency response of the device so that the instrument will not respond easily to signals outside the audible frequency range. A-weighted readings are used to estimate the probability of hearing damage in industry. These values have also been correlated with the annoyance caused by traffic and aircraft noise. The B-weighting network is rarely used. The Flat response of the meter is employed as a calibrated input to a frequency-band filter set or tape recorder.

The sound level meter weighting networks were chosen by the American Standards Institute as the reference curves for sound loudness contours for the three frequency responses available in the instrument. In general, these devices are not well suited for measuring intermittent noises that vary rapidly with time. In such cases, high-speed level recorders (graphic level recorders with high-speed pens) are employed. Such devices can record transient sounds whose sound-pressure level rises 25 dB within 25 msec.

Analyzers are devices that provide information on the distribution of the noise signal energy as a function of its frequency. For continuous noises, the meter's signal is first recorded in the field on a magnetic tape. The noise is then reproduced and analyzed in the laboratory. Analysis with octave filters provides the necessary information for acceptable levels.

A sound generator is used for measuring sound-transmission loss, absorption, and reverberation time. These parameters are evaluated in terms of discrete frequency bands or sinusoidal frequencies. Oscillators or random-noise generators are the most commonly used signal sources.

NOISE REDUCTION

Noise transmission is greatest in unlined metal ducts. The amount of noise reduction occurring in such designs is almost negligible (on the order of 0.1 dB of sound reduction per foot of duct even at frequencies greater than 1,000 Hz). Whenever economically feasible, sound-absorbent materials should be introduced in a duct as lining. Sound-attenuation calculations become complicated as large numbers of physical data become necessary. Information on acoustic resistance, acoustic reactance, and the phase angle between the two are required. Empirical correlations exist and calculation procedures have been developed for determining noise-level reductions in ducts lined with sound-absorbent materials.

It is often good design to install a traverse baffle or a sharp-angled bend in the channel. As sound travels down a lined duct, it becomes more highly absorbed around the perimeter than in the center. This results in a sound-pressure loss at the duct boundaries caused by destructive interference between the boundary and direct-reflected sound at grazing incidence. A sound-absorbent baffle can reduce the sound energy concentration near the center of the duct by causing both absorption and reflection.

The use of package attenuators has proven successful in air-conditioning noise reduction. Attenuators are short duct inserts whose cross section is comprised of zigzag paths. These paths achieve high acoustic absorption without causing large frictional resistance to air flow. Many designs are available commercially.

In designs requiring a large number of ducts that are supplied by one main fan, plenum chambers should be employed. A plenum chamber is a muffler, the interior of which is lined with sound-absorbent material. Figure 9-4A illustrates a single plenum chamber (note that inlet and exit ports are never located directly opposite each other).

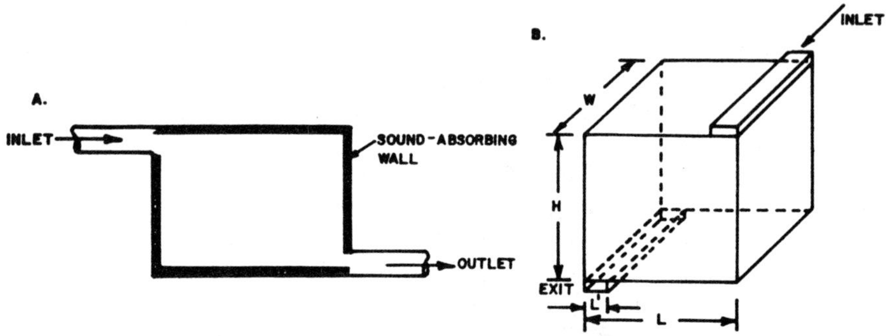

Figure 9-4 (A). Single-plenum chamber design; (B). three-dimensional illustration of a single plenum.

Figure 9-4B illustrates the geometry of a single plenum. The transmission loss associated with a plenum can be approximated by:

$$L_{TL} = -10 \log_{10} \left[A \left(\frac{\cos \theta}{2\pi d^2} + \frac{1-\alpha}{a} \right) \right]^2$$

where

A	=	cross-sectional area of exit port (ft^2)
	=	random-incidence absorption coefficient of the plenum lining (dimensionless)
a	=	total lined area in chamber times (ft^2)
d	=	distance between inlet and exit port (ft)
$\cos\theta$	=	H/d (where H is the height of chamber)
L_{TL}	=	transmission loss (dB)

The preceding equation can be used at high frequencies and small values of L'/L (Figure 9-4). At low frequencies, calculations underestimate actual transmission losses by 5 to 10 dB.

For greater sound reduction, multiple-plenum chambers can be employed (Figure 9-5). By knowing the transmission loss for a single chamber (L_{TLs}) and for a double chamber (L_{TLd}), the losses for a unit consisting of n chambers can be estimated from:

$$L_{TL_n} = (n-1)\, L_{TL_d} - (n-2)\, L_{TL_s}$$

where

$L_{TL\text{-}n}$ = transmission loss for a plenum with *n* number of chambers(dB)
n = number of chambers

Figure 9-5 Multiple-chamber plenum

Careful consideration should also be given to the individual components of the system. Proper selection of fans, unit coolers, and mountings should be done in the earliest stages of design. Table 9-2 lists some of the more common approaches to noise control of various units.

Engineering judgment must be used in following established guidelines and in selecting the proper NC curve for a particular design. Factors such as workers' attitudes toward noise levels and economics must be carefully weighed.

Table 9-2 Methods of Noise Reduction for Major Ventilation System

Unit	Noise-Control Technique
Compressors	Isolation from distribution network; should be mounted on resiliently supported inertia blocks
Condensers	Damp surfaces by applying a mastic compound
Cooling towers	Isolation from distribution network; towers should be mounted on heavy-duty ribbed neoprene pads to reduce vibrations
Evaporators	Evaporation pipe between the capillary exit and the evaporator entrance should be enclosed with a heavy flexible tubing
Fans	Isolation; generally located in a plenum
Piping	Pipe connections should consist of flexible couplings and pipes supported by resilient hangers
Pumps	Isolation from distribution network and mounting on resilient pads

10 INSTRUMENTATION

Noise measurements usually fall into one of two categories. The first category consists of sound level measurements usually made of the dBA level, for example, OSHA surveys and product noise ratings. Common measuring instruments are sound level meters and dosimeters. The second category consists of analyses made in support of environmental noise reduction programs and duct design programs. In addition to measurement of dBA level, such analyses require frequency analysis or amplitude distribution analysis.

MICROPHONES

Microphones are an essential part of all acoustical measurements, and microphone characteristics control the sections of sound level meter specifications and ANSI standards that deal with such parameters as frequency range, directivity, temperature stability, and effects of humidity. In fact, the microphone is the main distinguishing factor among different types of sound level meters.

Ideally, the microphone should produce an electrical signal that is an exact replica of the acoustical disturbance. It must operate over a wide dynamic range and a wide frequency range, and it must be stable under severe changes in environmental conditions. Of the many different principles of microphone construction, only condenser and piezoelectric microphones are used for instrumentation purposes. The condenser microphone can be of the stretched metal diaphragm or the electrode type. The stretched metal diaphragm condenser microphone is the best choice for accurate and repeatable measurements, while the piezoelectric and electrode condensers are the next best choice if price is important.

The construction of a stretched metal diaphragm microphone, commonly called a condenser microphone, is shown in Figure 10-1. The microphone cartridge consists of a thin metallic diaphragm in close proximity to a rigid backplate. These two elements are electrically insulated from each other and form the plates of a capacitor. A DC polarizing voltage is applied across plates.

Variations in pressure due to sound waves will move the diaphragm, thus varying the width of the air gap.

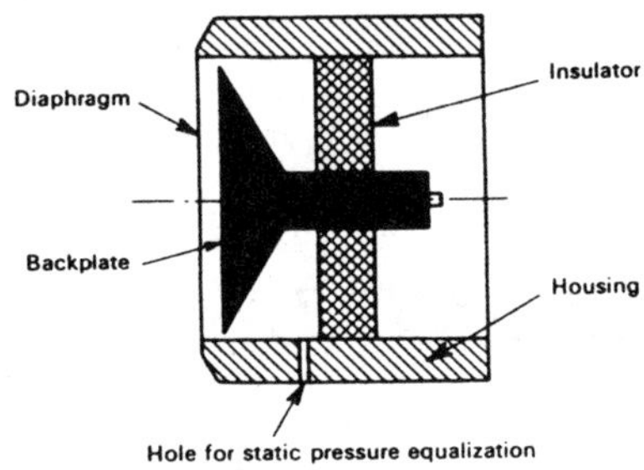

A

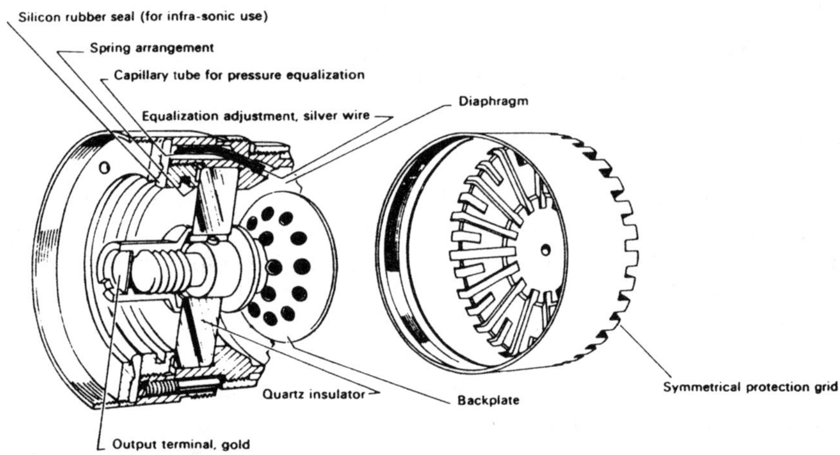

B

Figure 10-1 (A). Schematic construction of the condenser microphone.
(B). Construction of a condenser microphone.

Consequently, an alternating charge is generated on the capacitor. By careful design it is possible to keep the electrical output proportional to the sound pressure over a wide range of frequencies and dBA levels. In order that changes in atmospheric pressure will not change the static position of the diaphragm and cause a change in microphone sensitivity, the air inside the microphone is vented to the atmosphere. Therefore, the static pressure remains the same on both sides of the diaphragm. A common temperature coefficient for the metals in the diaphragm and housing is selected so that temperature stability of the microphone can be guaranteed for long periods.

There is another type of condenser microphone, which is commonly called the Electrode microphone. Its schematic construction, shown in Figure 10-2, is similar to a stretched metal diaphragm condenser microphone in that it has

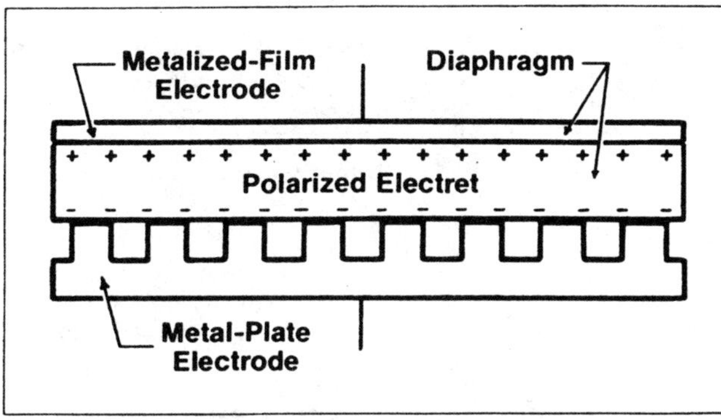

Figure 10-2 Schematic diagram of Electrode microphone.

a diaphragm separated from the backplate by a thin air gap. The diaphragm of an electrode microphone, however, is made up of a polymer film whose outside surface is metal plated. The plating and the backplate form the electrodes of a condenser with each raised section of the metal plate electrode representing a miniature microphone. The outputs are all summed to produce the effective output of the microphone. Since the polymer has electrical charges embedded into it, thereby containing its own charge, no polarization voltage is required.

The most common form of a piezoelectric microphone uses a piezoelectric ceramic element to which a bending movement is applied when the diaphragm is exposed to sound pressure. An example of this type of microphone is shown in Figure 10-3. The ceramic bender element is supported at both ends. The conical diaphragm applies a force to the center of the bender, thus producing an output voltage proportional to the amplitude of the diaphragm motion.

Piezoelectric microphones are often used in general-purpose sound level meters. They are relatively inexpensive and have the advantage that they do not need a power supply. For precision sound level meters, however, piezoelectric microphones are generally not suitable. Size for size, piezoelectric microphones have poorer frequency response and lower sensitivity than condenser microphones.

An instrumentation microphone should operate in a sound field without affecting the progress of the sound wave. One definition describes sound pressure as the pressure that would exist at a point in space if a microphone were not present. Unfortunately, the 1-inch and 0.5-inch microphones commonly used on sound level meters do affect progress of the sound wave. The interference occurs at higher frequencies where the short acoustical wavelengths approach the diameter of the microphone.

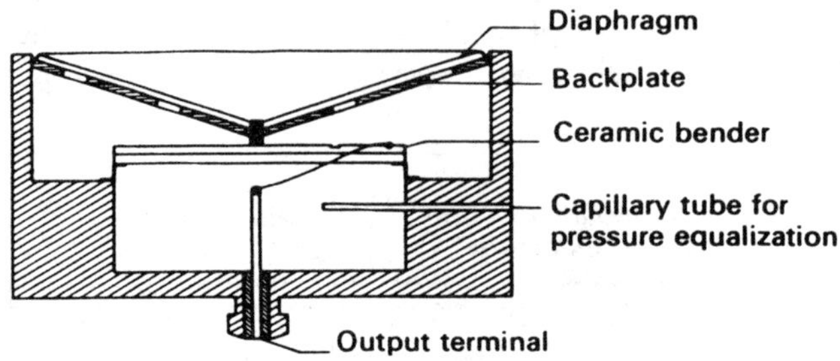

A

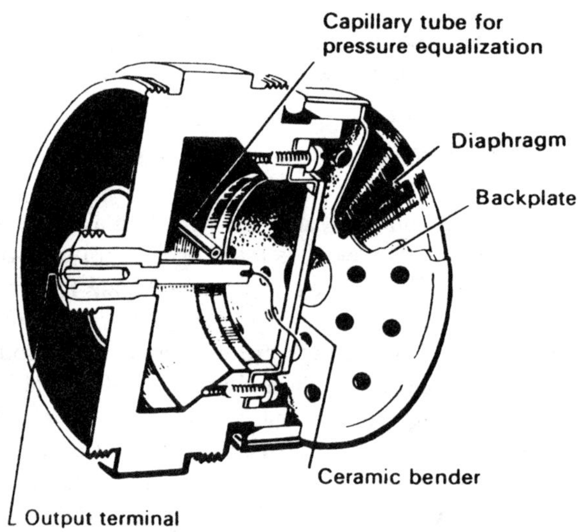

B

Figure 10-3 (A.) Schematic drawing of a piezoelectric microphone; (B.) Practical construction of a piezoelectric microphone.

There is a large additive effect when the sound wave is perpendicular to the plane of the microphone's sensing surface. For other angles of incidence, the sensitivity is lower. As a result, a microphone is omnidirectional at low and mid-frequencies, but it develops a large directional effect at higher frequencies. Figure 10-4 shows a typical family of corrections to the pressure-response curves of so-called random incidence, or grazing incidence, microphones. (A pressure field is one in which there is no wave propagation.) The name of this class of microphone is derived from its flat frequency response in the random sound field of a reverberant room and in a free field where the microphone is oriented so that the sound wave grazes across the face of the microphone.

However, in factory and community environments where more than a single sound source is present, sound waves more perpendicular to the

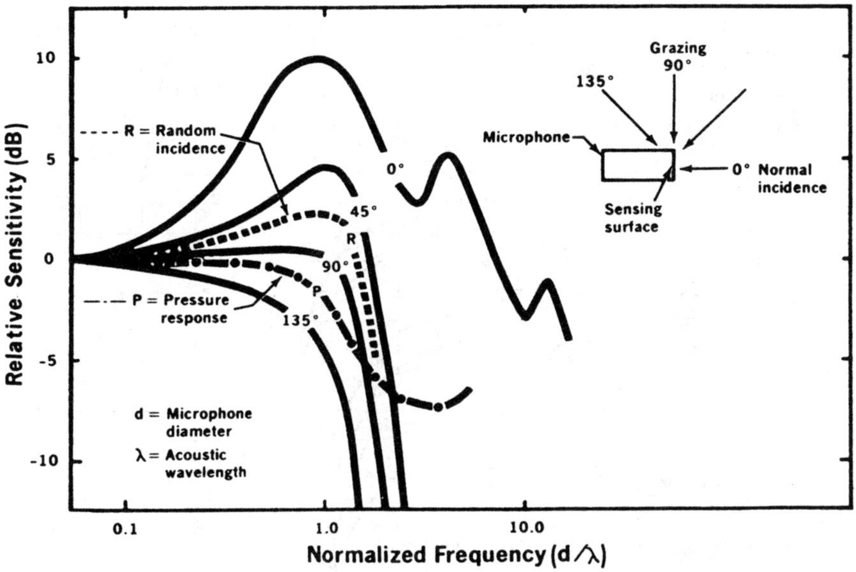

Figure 10-4 Effect of microphone interference on the sound field.

diaphragm may be heavily amplified and those coming from behind the microphone may be attenuated. Maximum amplification occurs when microphone diameter equals the wavelength of sound. For a 1-inch microphone this occurs at about 13.6 kHz. Because the pressure response begins to roll off at a somewhat lower frequency, the actual directional response curves peak out at 5 to 8 kHz, depending on the design of the individual microphone. So the potential for overestimating the sound level with this class of microphone lies in the range of 1 to 8 kHz. Unfortunately, this is the frequency range emphasized by A-weighting filters used in the measurement of dBA.

An alternate microphone is the so-called free-field or perpendicular incidence microphone. It is designed to have an overdamped pressure response curve in order to equalize the perpendicular incidence error discussed previously. This class of microphone has a flat frequency response for sound waves that are perpendicular to the plane of the sensing surface. Hence, its advertised axis of response is also its most sensitive axis. It has the advantage that it can never overestimate sound level, regardless of angle of incidence. In fact, the microphone can be oriented for maximum reading in order to get the most accurate measurement of sound level. A partial family of directivity curves for a grazing (90°) incidence and a perpendicular (0°) incidence microphone are shown in Figure 10-5. In many designs of microphones, the directional effects are even greater.

Because many environments consist of many intermittent sound sources, maximum accuracy is achieved if the microphone is truly omnidirectional over the entire frequency range of interest. This can be achieved by using a tiny

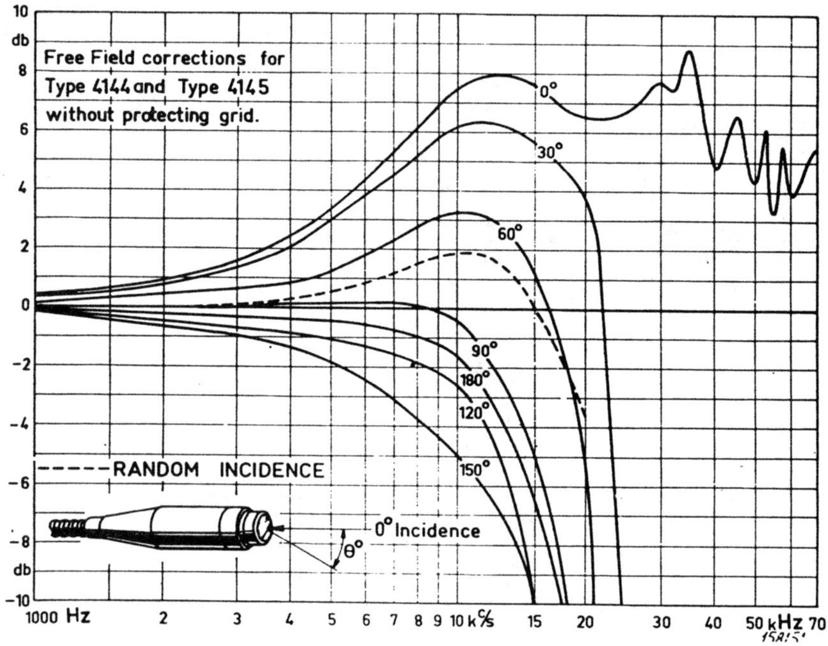

Figure 10-5 Free-field corrections for microphone without protecting grid (electrostatic actuator method of pressure calibration).

microphone, but the reduced sensitivity of smaller microphones makes this solution impractical. An alternative is to use adapters that eliminate directional characteristics by making the microphone appear to be smaller than it actually is.

Microphones are often used with wind screens. Although designed to reduce the effect of wind noise for measurements outdoors, they also serve to protect the microphone from oil spray, flying chips, and dust during factory noise measurements.

Good quality microphones will be stable over the period of measurement, but the data should be backed up by an acoustical calibration before and after each day's use. Humidity is seldom a problem, with the provision that moisture must not condense on the sensing surface.

SOUND LEVEL METERS

A sound level meter is an acoustical measuring instrument consisting of a microphone, amplifier, weighting filters, and readout meter, as shown in Figure 10-6. It will meet one or more governing standards: ANSI S1.4-1971 Specification for Sound Level Meters dominates in the United States, but IEC-179, an international specification for a precision sound level meter, is often

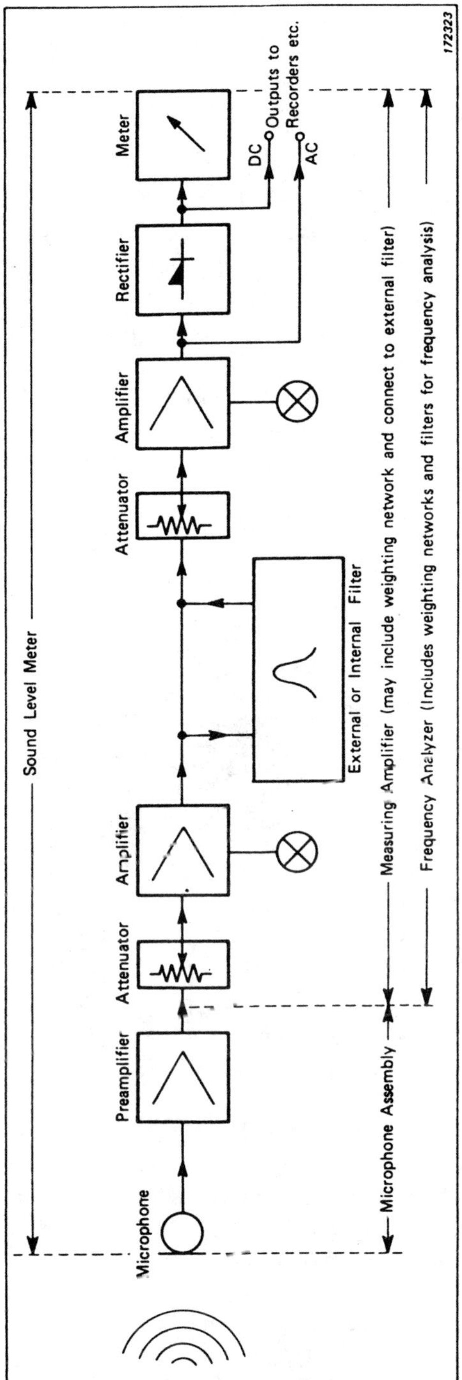

Figure 10-6 Block diagram of the important elements of a sound level meter.

specified for more exacting work. S1.4-1971 specifies three types of sound level meters:

Type 1 precision sound level meter
Type 2 general-purpose sound level meter
Type 3 survey sound level meter

The standards outline straightforward electrical specifications such as meter damping, response of frequency weighting networks, and certain environmental characteristics. But the distinguishing factor among the three types of sound level meters is the tolerance allowed on frequency response and directional response. Therefore, the microphone determines whether a sound level meter meets Type 1, 2, or 3 requirements.

The A-weighted frequency-response tolerances of current standards are summarized in Figure 10-7 for the important range of about 1 kHz. Note that precision sound level meters can have no response above 15 kHz, and general-

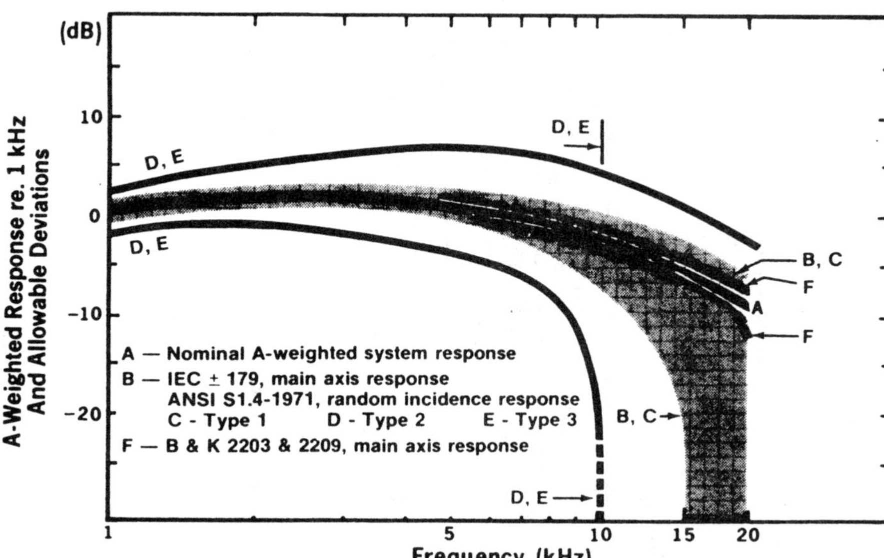

Figure 10-7 Current standards permit wide tolerances in A-weighted frequency response of sound level meters. However, close adherence of an instrument to the "A" curve providesprotection against severe amplification or attenuation of the data obtained in using the instrument in noise-measurement programs.

purpose and survey sound level meters are guaranteed to operate only up to 10 kHz. Most users demand more than the guaranteed minimum, so the prospective user must check the specifications for individual sound level meters.

Of much greater importance to the user are the directional tolerances shown in Figure 10-8. These curves are normalized so that conformance to the A-filter by main-axis measurements under free field conditions is represented by a flat response. The ANSI standards permit use of grazing incidence microphones, which overestimate sound waves with perpendicular incidence to the sensing surface. The Type 3 sound level meter is permitted to have positive

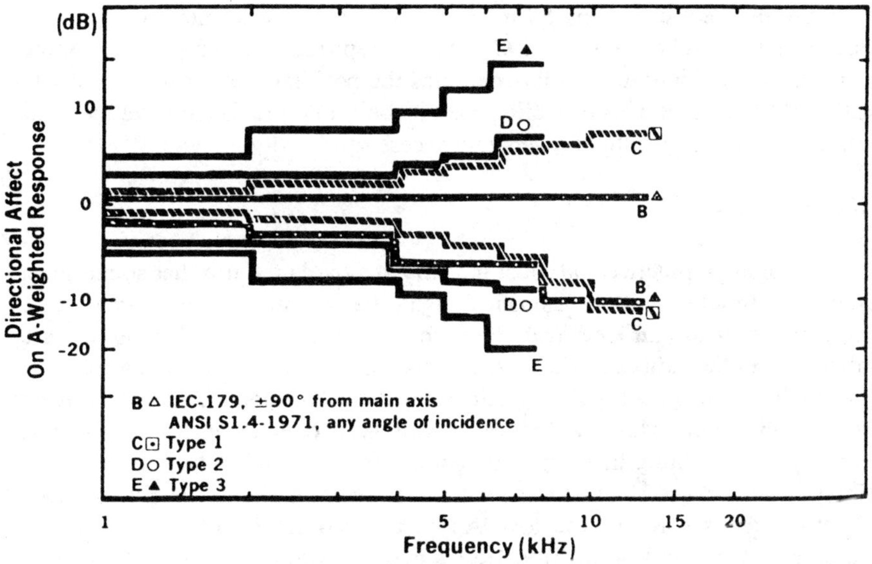

Figure 10-8 The standards permit irregularities in the directional characteristics of sound level meters. Industrial noise measurements generally require microphones that are more omnidirectional than some standards guarantee.

errors ranging up to 14 dB at 8 kHz. To protect the user, OSHA has disapproved the use of Type 3 meters for operator exposure surveys. The tolerances for Type 2 meters are much tighter.

Tolerances for Type 1 meters are similar to Type 2, but are extended to 15 kHz. Many organizations are not satisfied with Type 1 directional tolerances so the IEC-179 standard is often specified. The IEC-179 standard essentially requires that the most sensitive axis coincide with the main axis response. The intent is to specify a perpendicular-incidence microphone that can never overestimate sound level.

One other distinction in sound level meters should be made. If the meter is to be used for noise-reduction programs, a meter should be selected that has optional octave filters available. These meters are generally more expensive than those designed solely for noise surveys.

Sound level meters measure the rms level of the acoustical signal. Two meter damping positions are required—"fast" and "slow"—corresponding to averaging times of approximately 200 and 800 ms. Even in slow response the meter will often fluctuate. In this case the operator must note the range of levels indicated on the meter. For transient noise the operator notes the maximum reading during the event. But OSHA imposes a measurement requirement for impact noise that is not covered in the ANSI and IEC standards. For impact noise such as forging, stamping, and press operations, an OSHA noise limit is imposed based on the peak sound-pressure level. A peak-responding detector with a rise time not exceeding 50 microseconds is needed to measure this, and measurements must be made in the C or linear response mode. Furthermore, the meter must have a hold-circuit that captures the peak level and decays at a rate less than 0.05 dB/sec. This capability may be built into the sound level meter or the manufacturer may supply a separate accessory.

CALIBRATION

Common practice and most industry standards require that sound level meters be calibrated with an acoustical signal before and after each day's use. Most calibrators use an electrical signal to drive a diaphragm that serves as a loudspeaker in the calibration cavity. Because the calibration level is a function of the applied voltage, a regulating circuit is used to maintain the supply voltage at a constant level. Most calibrators of this type operate at 1 kHz. At this frequency the weighting filters have no gain, offering the advantage that a 1-kHz calibrator can be used with the sound level meter in the A-weighting mode without using any correction factors. A typical electrically controlled acoustical calibrator is shown in Figure 10-9. Here a piezoelectric ceramic element is used to drive the metallic diaphragm, which acts as a loudspeaker.

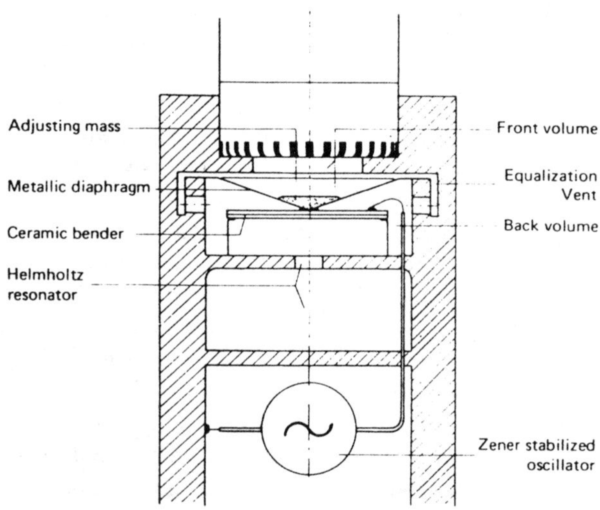

Figure 10-9 Cut-away view showing principle of operation of a portable acoustical calibrator.

THE ACOUSTICAL MEASUREMENT

Today's acoustical measurements are made either to evaluate an environment or to measure the noise output of a machine. Generally, there are several noise sources in the area, and the operator's machine may not be the major noise source. It is important that the microphone for measuring operator environment be omnidirectional—that it accurately measures all sound sources, regardless of their direction from the operator.

In contrast to operator position measurements, machine measurements call for measuring only the noise emanating from a single source. Machine measurements must usually be made on the factory floor, particularly frequency analyses in support of enclosure design. In these cases, measurements must be made close to the machine to ensure that the machine sound dominates the environment. This measurement is best made when all other machines are shut down, such as during nonworking periods, so that the most accurate characterization of the machine can be made. With the exception of dose measurements for which mobile employees wear a microphone on their lapel, the person making the measurement should be behind the microphone so that his or her own body does not act as a reflecting object to disrupt the measurement.

DOSIMETERS

The OSHA law requires characterization of varying noise environments by a single number criteria. The environment may be that experienced by mobile employees whose work takes them into many different factory areas, or it may be a fixed work area where intermittent machine operations produce unpredictable variations in noise level. OSHA specifies a maximum daily noise dose (D) of unity:

$$D = \frac{C_1}{T_1} + \frac{C_2}{T_2} + \cdots + \frac{C_n}{T_n}$$

where C is the total exposure at a given steady dBA level and T is the maximum allowable exposure time at that level during an 8-hour work day. The relationship between dBA level and allowable duration is shown in Figure 10-10.

Noise dose can be predicted for mobile workers by studying their movements and estimating the total time they are exposed to each different dBA level they experience. Alternatively, dosimeters can be worn by mobile employees and by the stationary employees to obtain a direct measure of noise dose. Dosimeters are integrating sound level meters that operate over fixed time periods, usually 8 hours. They convert dBA level to frequency in accordance with the OSHA dBA/exposure time relationship and produce a composite exposure number on a digital readout. Their advantage is that they are simple instruments that convert time varying sound level to an equivalent steady level. A typical OSHA dosimeter block diagram is shown in Figure 10-11.

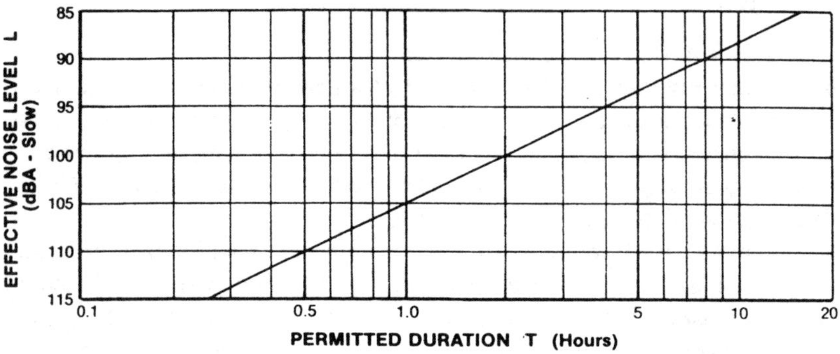

Figure 10-10 OSHA (noise) table of dBA versus allowable exposure for working personnel.

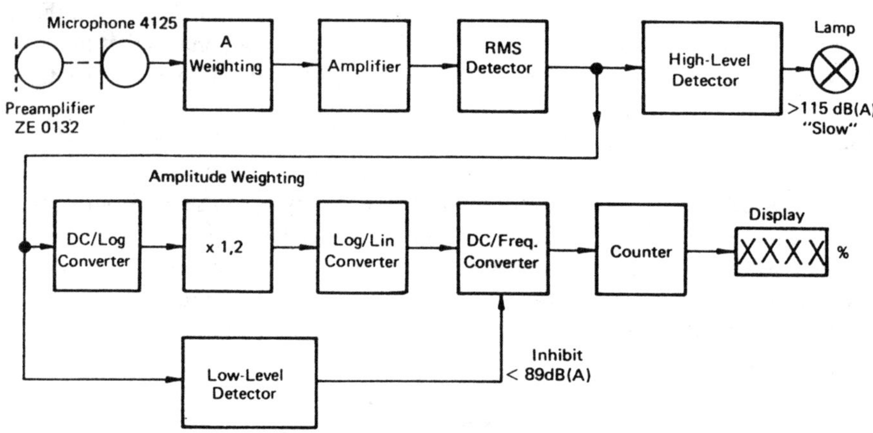

Figure 10-11 Simplified block diagram of a personal noise dose meter, complying with OSHA criteria.

FREQUENCY ANALYZERS

Basic sound level meters are commonly used to determine if there is an OSHA violation of steady noise exposure or if the overall noise level of a machine meets its guarantee. As such, they are go/no = go monitors. The readings give little assistance to the person responsible for reducing noise levels. The responsible person must usually know the frequency content of the sound because the design of efficient barriers and enclosures heavily depends on the frequency spectrum, and absorbing materials on walls or ceilings only work if high frequencies dominate. Because sound level meters are portable instruments, portable filters are also required. Filters take the form of an octave or a 1/3-octave sequence of bandwidths that permit the wide range to be reduced to a practical number of segments. The center frequencies and band-pass characteristics of octave and 1/3-octave filters are controlled by ANS1 S111-1966. A set of octave response curves is shown in Figure 10-12. Although analyzers used in the laboratory contain a large number of individual filters operating in parallel, portable analyzers consist of a single filter that the operator tunes or steps from one position to the next.

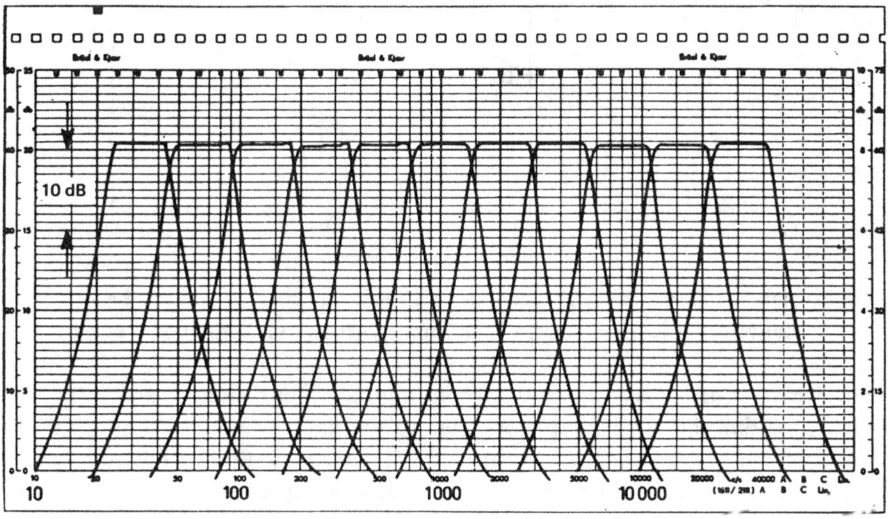

Figure 10-12 Graphic recording of the frequency characteristics of an octave filter set.

Historically, sound level meters have been designed so that octave analyses can be made only when the meter is in its flat frequency-response mode, thus preventing analysis of A-weighted signals. But today almost all noise reduction tasks are directed toward reducing dBA level. Therefore, it is necessary that today's instruments permit series connection of the A-weighting filter and the octave filters. In this way, the dominant A-weighted octave band is directly identified. A typical octave analysis is shown in Figure 10-13. The dBA level is plotted in octave bands to show the beneficial effects of lining the enclosure with sound-absorbing materials.

AMPLITUDE DISTRIBUTION ANALYZERS

As octave analyzers supplement sound level meters in programs to reduce dBA level, the dosimeter also needs a support instrument to identify dominant machines. Management needs an economical solution to noise problems; therefore, dominant noise components must be identified before any noise-reduction program is initiated. The machine that is the noisiest one in the shop when it operates is not necessarily the dominant noise source in a mixed

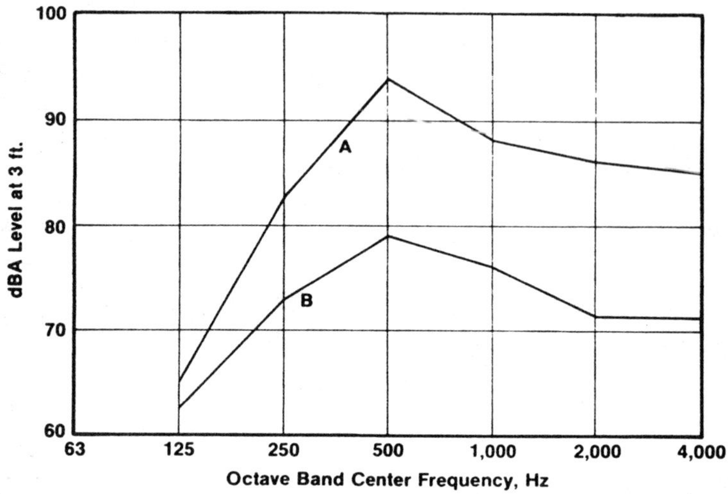

Figure 10-13 Noise level 3 feet from an enclosed electric motor. Design A is a standard enclosure; design B is lined with sound-absorbing material.

noise environment. Its contribution can only be determined by integrating the total time that the machine operates during the day. And it is inappropriate to build an enclosure for a noisy machine if an equal noise reduction can be achieved by reducing background noise, such as by conveyors or metal tote pans.

A typical instrument can be used to identify dominant noise sources. It classifies dBA levels into 12 dBA bands over the OSHA noise range and totals the accumulated noise level in each amplitude band.

One application of the Environmental Noise Classifier is in the measurement of the OSHA mixed exposure level where it serves as an area dosimeter and as a guide to the most economical solution to noise-reduction problems. Consider the data shown in Figure 10-14. The coefficient values are the ratios of the time spent in each band divided by the maximum allowable time in the band during an 8-hour day. If the sum of all coefficients (mixed exposure level) exceeds unity, the noise is in violation of the OSHA code. In each example, the times spent in the individual bands are read from the Environmental Noise Classifier in minutes and recorded in the table in hours. In each example, the mixed exposure level is about 1.25.

Without going into great detail, it can be seen that quieting the obvious machine that operates for one-half hour at the 97 to 100 dBA level in Case 1 is not an economical solution to the problem. If reduced to below 90 dBA, the OSHA coefficient would still be 1.00. Furthermore, quieting a large machine by 7 to 10 dBA is a major undertaking and may require construction of costly enclosures. The enclosures create problems of their own, such as access, increased labor, and temperature build-up. The data suggest the reduction of ambient noise and lower level machine noise to accomplish the goal.

dBA LEVEL	MAXIMUM HRS/DAY	CASE 1		CASE 2	
		ACTUAL HOURS	OSHA COEFFICIENT	ACTUAL HOURS	OSHA COEFFICIENT
< 90	–	2.5	–	–	–
90 – 92	6	3	0.5	1	0.17
92 – 95	4	2	0.5	1	0.25
95 – 97	3	0	0	2.4	0.80
97 – 100	2	0.5	0.25	0	0
OSHA MIXED EXPOSURE LEVEL			1.25		1.22

Figure 10-14 Example of amplitude analysis of OSHA environments.

In Case 2, there is no choice but to work on the noisiest machine. But the histogram suggests how many decibels the level must be reduced. In order to reduce the mixed exposure level to less than unity, the offending machine must be reduced to the 90 to 92 dBA band, or a reduction of 3 to 5 dBA. The same reasoning can be applied to reducing the noise of machine tools where the sound level changes with each cycle.

11 AUDIOMETRIC TESTING AND DOSIMETERS

AUDIOMETERS

The periodic testing of the hearing ability of persons is called audiometry. In industry, such testing may be carried out by a physician, qualified technical personnel, or anyone who will use techniques that meet the minimum requirements of OSHA. The major instrument used in audiometric testing is the audiometer, which measures a person's hearing ability. To comply with OSHA, audiometers must meet the specifications for limited-range, pure tone audiometers as set forth in ANSI Standard S3.6-1969, which is titled *Specifications for Audiometers*.

When measuring the threshold hearing levels of industrial personnel, the limited-range audiometer is most suitable. Under the ANSI specifications, limited-range audiometers must be capable of producing tones of the following frequencies: 500, 1,000, 2,000, 3,000, 4,000 and 6,000 Hz from 10 dB to 70 dB. This must be done for air conduction tests, and bone conduction may be neglected.

The ANSI Standard does not give preference to either automatic or manual audiometers; it permits the use of either. An operator is required to be present throughout the audiometric test when a manual audiometer is being used because the operator must manually present the various intensities at specific frequencies. The operator will establish the threshold levels at each frequency and will manually record them by means of the subject's response or lack of response.

When automatic audiometers are used, the test subjects will respond to the intensity of the tones given to them by releasing or pressing an electric switch. When there is no response, the audiometer will increase the intensity of the signal until there is a response, and then the intensity will decrease. This process automatically records the subject's threshold levels as they are obtained. With the automatic audiometer an operator is only needed to explain the test to the subject, position the earphones, and press the starting button. Table 11-1 gives the important aspects of both types of audiometers.

Table 11-1 Distinguishing Characteristics of Manual and Automatic Audiometers

	Relative Merits	
Factors Considered	**Manual Audiometer**	**Automatic Audiometer**
Time of test	Depends somewhat upon operator skill.	Usually 30 sec per frequency for each ear automatically controlled (6-7 minutes per subject).
Operator's time required	Operator must explain test, fit earphones, present tones at various levels of intensity at each frequency, and record thresholds. Trial period may be given before test to be sure subject understands.	After explanation and fitting of earphones, operator pushes start button. Instrument automatically presents tones, varies intensity and frequency and records thresholds. Trial period may also be given.
Threshold judgments	Made by operator based on subject response.	Recorded by instrument based on subject response.
Simultaneous testing of more than one person	Only one person can be tested at a time.	Limited only by number of audiometers and/or testing booths; however, four or five subject limit is practical
Retest	Operator can retest a particular frequency immediately should malingering or inattentiveness be suspect.	Retest can be given at a particular frequency after complete test, or by overriding the automatic if operator has test under surveillance.
Other	Results affected by operator fatigue and possible recording errors.	Instrument is not biased. Uniformity of tests. Test-retest reliability; less operator training required.

ACCURACY

Before being used for testing purposes, an audiometer must be checked to be accurate in such qualities as frequencies, purity of tones, and sound-pressure levels. A standard audiometer should be calibrated at least once every year by an expert, and in accordance with OSHA it must be biologically checked at least once a month, or before each use if used less than once a month. This is accomplished by having a subject with a known hearing ability take an actual test, the results of which are then compared to previous tests for stability. The company's personnel can perform this test and the results should be presented for possible inspections.

The audiometer room or booth in which the subject is placed during testing should be inspected regularly for damaged gaskets and seals. Also, sound level readings should be made occasionally to determine if the sound attenuation qualities of the structure are up to par.

In cases where audiometric tests are conducted outside the plant, the Department of Labor representative will inspect the facilities and test records. The plant management will make arrangements for such inspections with the person conducting the audiometric tests and may be present with the representative throughout the inspection and examination of the records.

RECORDS

There must be a statement on the audiometer that tells whether it is calibrated to ASA 1951 or to the current American National Standard. A log of all accuracy and biological checks of the audiometer must be kept available for inspection. Also, a record of each audiogram made on each worker must be filed and maintained for inspection. The records of all the audiometric tests will show whether readings are based on ASA 1951 or on the American National Standard, which is identical to the ISO 1964 Standard. The records of every worker required to be tested must be kept for one year after termination of employment, or after the worker has been transferred to an area with noise levels below 90 dBA.

All records shall be inspected for evidence of any deterioration of hearing ability and of the action taken to prevent further deterioration in the employees who have suffered some hearing impairment.

AUDIOMETRIC TEST BOOTHS

The site for the audiometric testing is important as the test tones can be masked easily by ambient noise. Unless an extremely quiet room is available, away from traffic, machines, people, conversations, and doors, an audiometric testing booth is required to conduct an accurate test.

In accordance with OSHA, the ambient noise level for conducting tests must meet the criteria as set forth in ANSI Standard S3.1-1960 (with latest revision), *Standard for Background Noise in Audiometric Rooms*. This standard lists the maximum allowable sound pressure levels of the ambient noise at each frequency of testing. Table 11-2 presents the allowable levels as given in the referenced standard.

Each location must first be checked for "quietness" with a noise-level meter utilizing either 1/3-, 1/2-, or full-octave band settings. For example, if the 1/2-octave band setting of a sound level meter is used for the range setting including 1,500 Hz, the allowable maximum ambient level cannot exceed 39 dB for audiometric testing. A sound level meter must be used with a frequency analyzer having 1/3-, 1/2-, or full-octave band settings. These can be borrowed or rented, and in some cases vendors of audiometric equipment will conduct a

Table 11-2 Maximum Allowable Sound Pressure Levels (dB Ref. 0.0002 microbar) for No Masking Above Zero Hearing Loss Setting of Audiometers

Test Frequency of Audiometer (Hz)	Octave Band	½-Octave Band	⅓-Octave Band
125	40	37	35
250	40	37	35
500	40	37	35
750	40	37	35
1000	40	37	35
1500	42	39	37
2000	47	44	42
3000	52	49	47
4000	57	54	52
6000	62	59	57
8000	67	64	62

site survey at no charge. If any table value(s) is exceeded, the site is not good enough for audiometric testing. Therefore, either the noise level must be reduced, an audiometric booth utilized, or another site selected. The test must be made when all machinery present is operating and noise level is at its highest.

One must remember that audiometric rooms or booths do not eliminate noise, they only attenuate it. The more attenuation required, the more material required for fabrication, and hence, the greater the cost. Audiometric rooms are available in all sizes, from smaller than telephone booths to 6 x 10 ft rooms and larger for multiple-person testing. Some booths are mounted on wheels or casters, while most are a permanent type that is too big to be moved through doorways. However, assembly can be performed by plant maintenance personnel, who must be sufficiently supervised, or suppliers of such units can provide installation crews at some cost.

As soon as the size of an audiometric room is established, the adequacy of its attenuation qualities can be determined by requesting attenuation data from the vendor before purchase. These data should contain the attenuation in dB of the booth or room at each frequency at which testing will be performed. A simple calculation can then be made for each test frequency as follows:

$$\frac{\text{Ambient noise level } (\in dB) + 10 = dB}{\text{Safety factor–booth attenuation}} = \frac{\text{Sound level inside}}{\text{audiometric room}}$$

This calculation can be repeated for each frequency.

A 10-dB safety factor is usually recommended to ensure a margin for any later increase in the ambient noise. The values calculated for sound pressure level at each frequency should then be compared with the maximum allowable values shown in Table 11-2. If no values are exceeded, the audiometric room is satisfactory for test purposes. If one or more values are exceeded, then the room is unsatisfactory at that specific site, and either the site must be changed, the noise abated, or another room selected with better attenuation qualities.

Rooms or booths can come with many options; standard features are usually one window, a light, and connections for the audiometer in addition to a door, walls, ceiling, and floor. Costs vary directly in proportion to attenuation and size. Extras range from an outside shelf for the audiometer to fluorescent lighting to rugs on the floor. However, two options that should be considered are vibration isolator rails to attenuate vibration transmitted through the ground and a blower-ventilation system.

ALTERNATE PROGRAMS

An alternative to setting up one's own audiometric program is to design a program with either a hospital or clinic or a mobile laboratory company. The cost of any hospital or clinic arrangement usually depends on the number of persons requiring testing. This may be the best method for companies with a small number of employees to test. However, employee time lost must be considered because the employees must leave the plant, travel to and from the hospital or clinic, and perhaps even wait to be tested.

Several companies offer in-plant testing of employees by means of mobile vans equipped with audiometers and booths. The vans come to the plant site, take sound level measurements to certify the site, and then give the employees the test, often at a rate of 20 subjects per hour, utilizing multiple instruments. Some companies require a minimum of 25 to 50 people to be tested before they will send out their van.

The mobile van companies will do all the work, qualify the site, provide the technicians, maintain records, submit reports, and analyze the test results. All the company has to do is to schedule the people to be tested. However, certain problems often arise. For example, some people might be absent the days the van is on site. Also, there is the problem of testing shift workers, and those people joining the company after the van has left. All those missing the test can be referred to a hospital or clinic. Added to this is the problem of "minimum time away from noise" prior to taking an audiometric test. A hearing test taken after exposure to noise can result in a temporary or apparent threshold shift and may appear to be a hearing loss or a permanent threshold shift. Authorities differ on the precise amount of rest time from noise before testing, but most agree on at least 16 hours. If followed, this presents a serious scheduling problem, particularly if utilizing a mobile van. And the workers should be kept away from noise that day until the time of their test.

If a company has its own equipment, it can schedule a few tests each morning before the person has entered the noisy area for work. This is even more feasible if the company will be utilizing existing employees to supervise the testing (such as nurses, medical technicians, personnel department people) as testing will only take up a small portion of their day. Such a program distributes the testing load more evenly throughout the working year.

DOSIMETERS

Dosimeters are portable instruments that can be attached to workers and go with them throughout their normal working tasks. The dosimeter measures and records the full noise exposure of the employees and at the end of the work shift gives the percentage of allowable noise exposure received by them. In effect, it computes the dosage/time equation for the entire working day. This equation is as follows:

$$\frac{C_1}{T_1} + \frac{C_2}{T_2} + \frac{C_3}{T_3} + \cdots \frac{C_n}{T_n}$$

where C_n is the total time of exposure at noise level n, and T_n is the total time permitted at that particular noise level. If the sum of all of these fractions is greater than one, then the combination of exposures has exceeded the permissible limit. Dosimeters are slightly larger than a pack of cigarettes and weigh about 8 ounces. Powered by two small batteries, they are completely portable and can be easily clipped to a shirt pocket or a belt.

Dosimeters contain a microphone for sound pickup, and some models use a built-in microphone. Various models have a microphone attached to the dosimeter through a lead-in wire, which makes it possible for the microphone to be placed anywhere on the employee. Therefore, the microphone can be placed in the ear like a hearing aid.

There are many types of dosimeters offered on the market. In choosing the best type for one's purposes, several factors should be taken into consideration. Individual variations should be carefully examined in the light of a company's overall OSHA compliance program, conservation program, and individual personnel policies. Another factor to consider is whether a dosimeter with an integral readout is needed. Dosimeters with integral readouts continuously integrate and display the percentage of daily noise exposure. These are self-contained and do not require the aid of any additional readout equipment.

An important factor in choosing dosimeters is economics. If many dosimeters are to be used continuously at one locale, the twoinstrument system can be less costly since only one readout indicator will handle any number of measurement units. However, this cost advantage must be weighed against the fact that operation is not as convenient as with the single-unit system. Also, additional maintenance is usually necessary to prevent the readout connections from becoming dirty or corroded, causing erroneous readings.

Another factor in comparing dosimeter costs is how the unit is powered. Some units come with a rechargeable nickel-cadmium battery that will last at least a year, providing a minimum of 8 hours use per charge. However, other dosimeters use disposable batteries that require frequent replacement.

12 INTERPOLATION AND MAPPING

Industrial noise is a common cause of hearing loss among workers. In addition, industry-generated noise that exceeds community ambient noise levels is a significant annoyance and may evoke community reaction.

NOISE LEVEL AND FREQUENCY WEIGHTING

Sound is characterized by its frequency and pressure level. Sound pressure level *LP* (decibels) is given by the logarithmic relationship

$$L_p = 10 \log_{10} (p^2 rms / p^2 ref) \tag{12-1}$$

where $p^2 rms$ is the mean square sound pressure (Newtons/m^2) and $P_{ref} = 2 \times 10^{-5}$. Alternately, intensity level L_I (decibels) is given by the logarithmic relationship

$$L_I = 10 \log_{10} (I / I_{ref}) \tag{12-2}$$

where I is the sound intensity in watts/m^2 and $I_{ref} = 10^{-12}$. For airborne sound and ordinary conditions, the difference between L_p and L_I is negligible.

Audible sound covers the frequency range from about 20 Hz to 20,000 Hz and ranges in sound pressure level from about $LP = 0$ dB (the threshold of hearing) to $L_p = 140$ dB or higher (the threshold of pain). The A-weighting network in a sound level meter combines the effect of sound at various frequencies approximately as the human ear responds. The A-weighted sound pressure level in decibels (dBA) is a convenient and widely used measure of industrial noise. For purposes of hearing conservation and assessment of community noise, measurements in dBA are generally adequate except when pure tones are prominent.

INDUSTRIAL HEARING DAMAGE RISK

Routine hearing-level tests are made with an audiometer, with hearing thresholds measured at several frequencies. Criteria for acceptable noise exposure depend on the definition of a hearing handicap. If a hearing handicap is defined as an average hearing loss of 25 dB, continuous (8 hours per day), industrial noise exposure to levels up to 80 dBA will not increase the chance of developing a handicap. This exposure level is based on the results of audiometry examinations of industrial workers and comparisons (by age group) with others who were not exposed to high industrial noise levels. Other research suggests a much lower safe exposure level on the basis of a different hearing impairment criterion.

COMMUNITY NOISE FROM INDUSTRIAL SOURCES

When industrial and residential zones abut, many problems can develop, the most pervasive often being noise. Residents may complain of sleep interference, speech interference, or simply annoyance. At a distance of 4 feet between talker and listener, speech communication is barely possible at normal male voice level with a background noise level of 63 dBA.

Sleep interference is more difficult to express quantitatively. In one study, it was found that the time required to fall asleep increased by 1 to 1.5 hours at noise levels between 50 and 60 dBA. Furthermore, at those noise levels, the depth of sleep was decreased. Annoyance is a very subjective effect. However, numerous studies have indicated that when intrusive noise exceeds background noise levels by 5 to 10 dBA, community action is likely.

NOISE CONTROL

Principal methods of industrial noise control include reduction of noise output at the source and control of noise transmission paths. Noise sources should be considered first. After reducing noise source output to the lowest practical levels, noise transmission paths through solids and air should be examined. For airborne noise outdoors, buildings and barriers between the noise source and receiver have a substantial effect. Finally, the distance between source and receiver may be increased, or if the receivers are industrial workers, they may be fitted with hearing protection.

COMPLIANCE WITH CODES AND STANDARDS

Basically, the Occupational Safety and Health Administration (OSHA) permits 8 hours per day exposure to noise levels of 90 dBA. The exposure time limit is halved for each 5 dBA increase up to the maximum permitted continuous noise level of 115 dBA. The allowable levels represent a compromise between actual existing noise levels in industry and safe levels in terms of hearing damage risk. Thus, OSHA levels may be subject to revision to reduce further the risk of hearing loss.

Community noise codes are generally based on annoyance and speech and sleep interference rather than hearing damage risk. Permitted levels for noise that intrudes on residential properties are much lower than permitted levels within factories. One city code, for example, has different requirements for light and heavy manufacturing zones. The combined output of a corporation in a restricted manufacturing zone is not to exceed 55 dBA at the boundary of a residential zone. The limit is 61 dBA for general manufacturing zones; for general and heavy manufacturing zones, measurements may be made at the zone boundary or 125 feet from the nearest property line of a plant or operation, whichever distance is greater. One state code limits continuous noise from industrial operations to 65 dBA between 6AM and 10 PM with a nighttime limit of 50 dBA. For this code, measurements are made at a residential property line (even if the residence does not lie in a residential zone). The preceding noise-limit values are given as illustrations only. The latest applicable codes and regulations must be obtained in every case.

Statistical (percent exceeded) noise levels are sometimes used to describe noise that varies with time. The most commonly used levels are L_{10}, L_{50}, and L_{90} values expressed in dBA and based on one-hour measurement periods. The L_{10} level is exceeded 10 percent of the time. For example, L_{10} = 65 dBA indicates that a noise level of 65 dBA was exceeded for 6 minutes during a given hour. Noise level L_{50} is the median, and L_{90} is the level exceeded 90 percent of a given time period.

NOISE-LEVEL MAPPING

The distribution of resources applied to noise control should be based on careful prediction, measurement, and planning. The plotting of noise contours (points of constant noise level) can be an important step in noise control. Using a few predicted or measured noise-level values, it is possible to estimate and plot noise levels at other locations. This permits the application of engineering controls to reduce noise transmission paths. Noise plots may also aid in siting of noise-producing operations to reduce noise impact on employees and on the neighboring community. Other administrative controls including stationing of employees and the requirements for personal hearing protection may be indicated by noise-level plots.

POINT SOURCES OF NOISE

A single piece of machinery, an exhaust stack, a fan, or other relatively concentrated noise source may be considered a point source. At locations sufficiently distant from the source, noise intensity follows the inverse square law, providing reflections are not present.

THE INVERSE SQUARE LAW

In the absence of barriers, an ideal point source generates a spherical wave. Noise intensity (watts/m^2) decreases with distance according to the inverse square law:

$$\frac{I_2}{I_1} = \frac{r_1^2}{r_2^2} \tag{12-3}$$

where intensity I_1 is measured at distance r_1 from the source and intensity I_2 at distance r_2.

Combining Equations 12-2 and 12-3, we obtain the effect of distance on sound level L(dBA) for an ideal point source:

$$L_2 = L_1 - 20 \log_{10}(r_2/r_1) \tag{12-4}$$

where L_1 and L_2 refer to sound levels at distances r_1 and r_2, respectively. When dealing with airborne sound under ordinary conditions, there is no need to distinguish between sound-pressure level and sound-intensity level. It can be seen that sound level due to an ideal point source decreases by about 6 dBA per doubling of distance from the source. This relationship holds for a point noise source on a hard floor as well.

THE NEAR AND REVERBERANT FIELD

No actual machine can be exactly represented by an ideal point source. The region close to an actual noise source is called the near field. Noise levels fluctuate within the *near field*, and the 6-dBA sound decrease per doubling of distance law (inverse square law) does not hold. The extent of the near field will be of the same order of magnitude as a characteristic dimension of the noise source. The *far field* is the region in which sound level decreases by 6 dBA per doubling of distance. Smooth walls that reflect sound waves are called *acoustically hard*. In a room, the sound level may fluctuate near walls. The region near a wall in which the 6 dBA per doubling of distance law does not hold is called the reverberant field. The near field, far field and *reverberant field* are illustrated in Figure 12-1.

Noise level is also influenced by barriers, by atmospheric attenuation, and by background noise. Barriers that interrupt (or nearly interrupt) the line of sight between the noise source and the receiver cause a substantial reduction in noise level. Atmospheric attenuation causes additional noise reduction when sound is transmitted for hundreds of feet. Background noise (traffic noise and noise

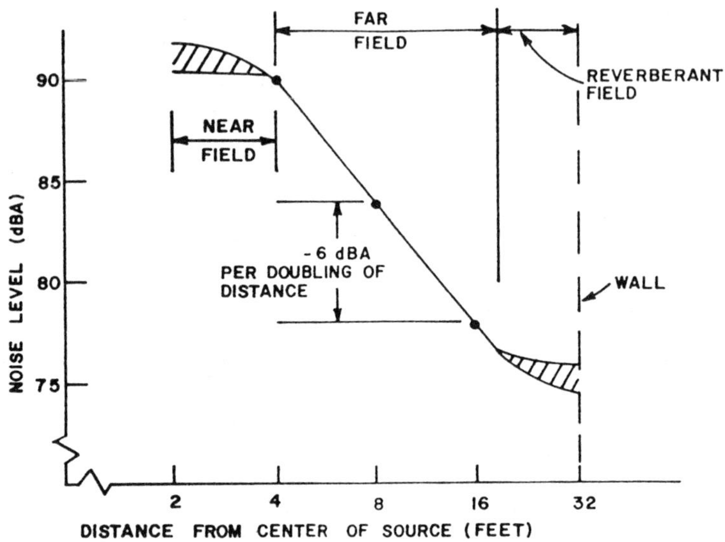

Figure 12-1 An illustration of the variation of noise level with distance from a point source (semilog coordinates).

noise from sources other than the machine in question) becomes prominent as the distance from the source increases.

PLOTTING POINT-SOURCE NOISE LEVELS

Using Equation 12-4, the difference in noise level ΔL (dBA) measured at distances r_1 and r_2 from a point source is given by:

$$\Delta L = L_2 - L_1 = -20 \log_{10} (r_2/r_1) \tag{12-5}$$

Conversely, for a given difference in noise level ΔL, the ratio of distances to a point source is given by:

$$r_2/r_1 = 10^{-\Delta L/20} \tag{12-6}$$

An increase in noise level, of course, corresponds to a decrease in distance. Using Equation 12-6, the relationship between changes in noise level and the ratio of distances from a point source is given in Table 12-1.

An interpolator for point-source noise may be constructed as shown in Figure 12-2. A vertical line of any convenient length is drawn at the left of the figure. The top of the line is marked 0 dBA and the bottom of the line represents the noise source location. Measuring from the bottom of the line, 0.891 times the length of the line corresponds to $\Delta L = 1$ dBA; 0.794 times the length of the line corresponds to $\Delta L = 2$ dBA, and so on, as given in Table 12-1. Lines are drawn from each of these points to a point of convergence at some convenient

Table 12-1 Change in Noise Level with Distance from a Point Source

Table 12-1 Change in Noise Level with Distance from a Point Source

Change in Noise Level ΔL (dBA)	Distance Ratio r_2/r_1
0	1.000
1	0.891
2	0.794
3	0.708
4	0.631
5	0.562
6	0.501
7	0.447
8	0.398
9	0.355
10	0.316
12	0.251
20	0.100

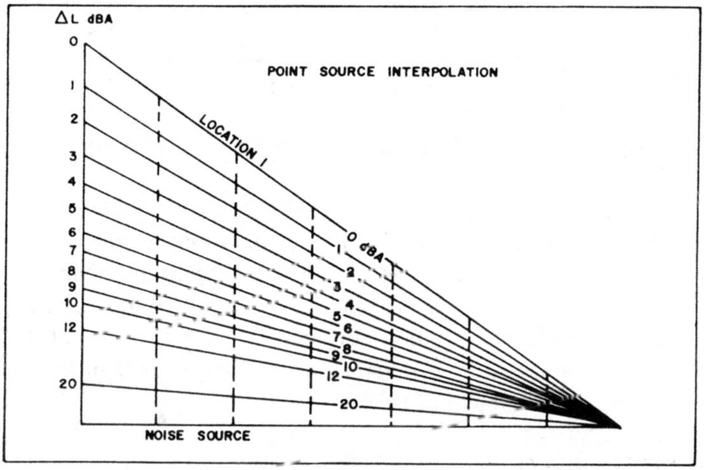

Figure 12-2 Point source noise interpolator.

location at the right of the figure. The interpolator of Figure 12-2 may be duplicated directly on a sheet of clear plastic, or a larger interpolator may be constructed from the data of Table 12-1 using transparency marking pens on clear plastic.

If the noise level is known at one point, the interpolator may be used to estimate the noise level at another point in line with the first point and the noise source. Using a scale drawing of the area, the interpolator is oriented so that the "location 1" line lies over the point farther from the source and the "noise source" line lies over the source. The line lying over the nearer point indicates the difference in noise level between the two points.

Figure 12-3 illustrates the use of the interpolator when the noise level is unknown at a point between the source and a point of known noise level. In this example, the interpolator gives $\Delta L = 3$ dBA, which is added to the known value of 70 dBA for a noise level of 73 dBA at the point nearer to the source. If the point of unknown noise level is farther from the source than the known point, ΔL is subtracted. If one of the points is very close to the noise source, it may lie in the near field, making the interpolation invalid. If one of the points is very far from the noise source, noise from sources other than the source in question may have a predominant effect.

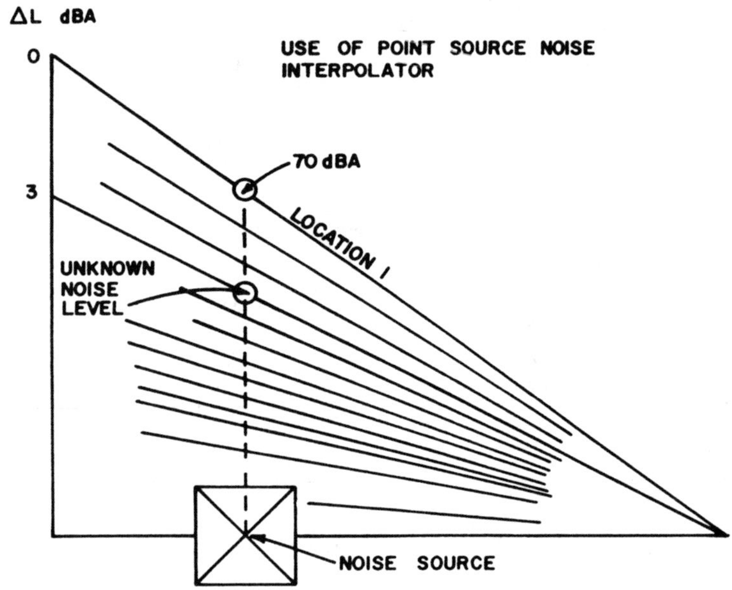

Figure 12-3 Use of point source noise interpolator.

LINE SOURCES OF NOISE

When a number of pieces of machinery or other noise sources of equal sound power lie in a line, the situation may be approximated by a line source. At points sufficiently distant from the line source, noise intensity follows the inverse distance law, providing reflections are not present.

THE INVERSE DISTANCE LAW

In the absence of barriers, an ideal line source generates a cylindrical wave. Noise intensity decreases with distance according to the inverse distance law:

$$\frac{I_2}{I_1} = \frac{r_1}{r_2} \qquad (12\text{-}7)$$

In terms of sound levels L_1 and L_2 (dBA) at distances r_1 and r_2, respectively, this becomes:

$$L_2 = L_1 - 10 \log_{10} (r_2/r_1) \qquad (12\text{-}8)$$

Thus, the sound level due to an ideal line source decreases by about 3 dBA per doubling of distance from the source.

PLOTTING LINE SOURCE NOISE LEVELS

Using Equation 12-8 for a given difference in noise levels, ΔL, the ratio of distances to a line source is given by:

$$r_2/r_1 = 10^{-\Delta L/10} \qquad (12\text{-}9)$$

Using Equation 12-9, the relationship between changes in noise level and the ratio of distances from a line source is given in Table 12-2.

Table 12-2 Change in Noise Level with Distance from a Line Source

Change in Noise Level ΔL (dBA)	Distance Ratio r_2/r_1
0	1.000
1	0.794
2	0.631
3	0.501
4	0.398
5	0.316
6	0.251
7	0.200
8	0.158
9	0.126
10	0.100

An interpolator for line source noise may be constructed as shown in Figure 12-4, using the data from Table 12-2. If the noise level is known at one point, the interpolator may be used to estimate noise level at another point if both points lie on a perpendicular to the line source. The interpolator is oriented so that the "noise source" line lies over the source and the "location 1" line over the farther point. The line lying over the nearer point indicates the difference in noise level between the two points.

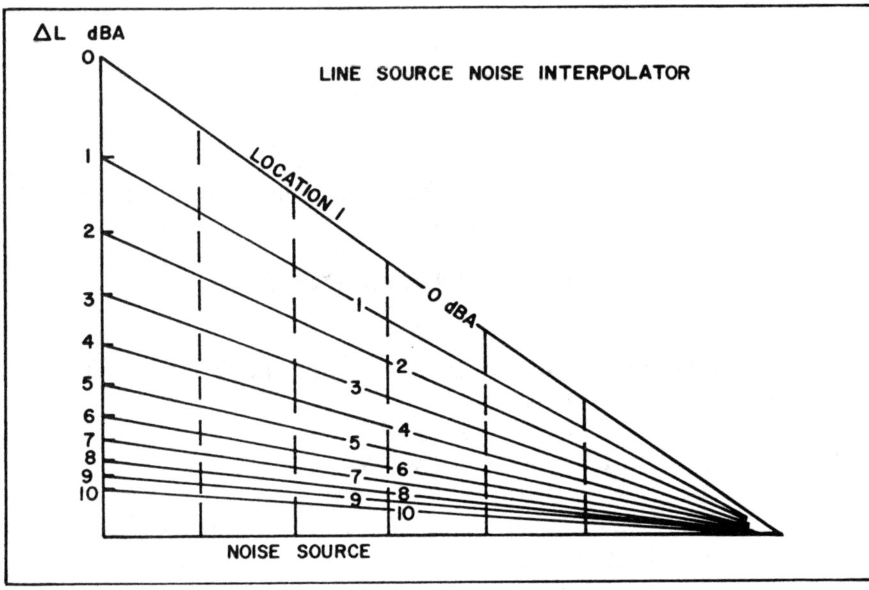

Figure 12-4 Line source noise interpolator.

COMBINED NOISE SOURCES

When two or more independent noise sources are present, the effect at a given point may be obtained by adding the noise intensities (watts/m^2). Most measurements and predictions, however, are given as noise levels L (dBA). For separate noise-level contributions L_1 and L_2, the total noise level L_T dBA at a point is:

$$L_T = 10 \log_{10} (10^{L_1/10} + 10^{L_2/10}) \tag{12-10}$$

It can be seen that for $L_1 = L_2$, $L_T = L_1 + 3$ dBA (approximately). In general, if L_1 is larger than L_2 by N dBA, the combined effect of L_1 and L_2 is larger than L_1 by:

$$L_T - L_1 = 10 \log_{10}(1 + 10^{-N/10}) \tag{12-11}$$

The difference is given in Table 12-3 for convenience in combining noise levels. The results are given in chart form in Figure 12-5.

Table 12-3 Combined Noise Levels

Difference in Noise Levels to be combined (L_1-L_2 dBA)	Amount Added to Larger Level to Obtain Total Noise (L_T-L_1 dBA)
0	3.01
1	2.54
2	2.12
3	1.76
4	1.46
5	1.19
6	0.97
7	0.79
8	0.64
9	0.51
10	0.41
15	0.14
20	0.04

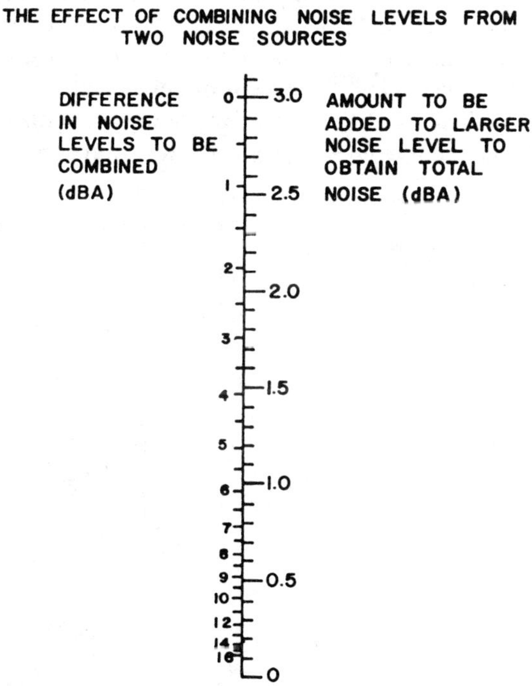

Figure 12-5 The effect of combining noise levels from two noise sources.

Using the table or chart, any number of noise contributions can be combined, two at a time. For example, let three machines produce levels of 70 dBA, 70 dBA, and 67 dBA at a given point when operating singly. The effect of the three operating together is given by combining the first two levels to obtain 73 dBA, and then combining 73 dBA and 67 dBA to obtain a total noise level of 74 dBA (approximately). Due to the difficulty in making precise noise measurements and the tendency of noise levels to vary with time, final results are generally rounded off to whole decibels. It can be seen that if two noise contributions are to be combined and one is 15 or 20 dBA below the other, the lower noise level will not have a significant effect. For example, 80 dBA combined with 60 dBA produces a total noise level of approximately 80 dBA.

NOISE CONTOUR MAPPING FOR LINE AND POINT SOURCES

It is sometimes necessary to map noise contours to estimate the impact of industrial noise on workers or on the neighboring community. The noise interpolators may be used as an aid in plotting in a free field, sufficiently far from the noise source, barriers, or reflecting surfaces. For an ideal line source, constant noise level contours are simply lines parallel to the source. For an ideal point source, constant noise level contours are concentric circles. For a source that shows some directionality, the contours will be distorted accordingly.

Figure 12-6 illustrates a plan for installing a row of machines that may be approximated by a line source. Based on measurements on similar machines, it is estimated that the row of machines will produce a noise level of 78 dBA at the points indicated. The line-source interpolator was used to draw lines of constant noise level parallel to the source as shown in the figure. A single

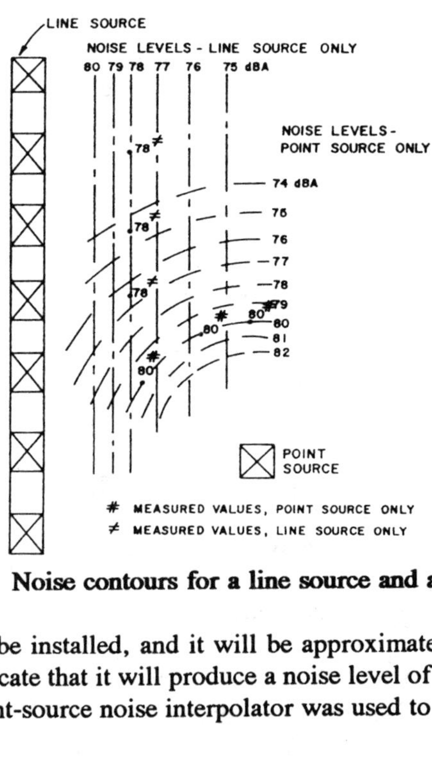

Figure 12-6 Noise contours for a line source and a point source.

machine will also be installed, and it will be approximated by a point source. Measurements indicate that it will produce a noise level of 80 dBA at the points indicated. The point-source noise interpolator was used to estimate the location of curves of constant noise level due to the single machine. No curves are drawn near the machines since the point-source and line-source idealizations are not valid in the near field. Barriers, reflecting walls, and ceilings must also be considered. In this example, it is assumed that the area of interest is free of obstructions.

NOISE CONTOURS FOR COMBINED NOISE SOURCES

Figure 12-6 illustrates the approximate effect of two noise sources operating separately. If both sources operate together, the combined effect will be of interest. This may be obtained as in Figure 12-7 considering the intersection points of the noise contours and using Table 12-3 (or Figure 12-5) to sum the noise contributions. For example, the intersection of two 77-dBA contours represents a combined effect of 77 + 3 = 80 dBA, and that point is so marked. The intersection of a 74-dBA contour and an 80-dBA contour represents a combined effect of 80 + 1 = 81 dBA (approximately) and it is so marked. As indicated previously, precision of better than 1 dBA may not be warranted.

The combined noise contours are obtained by joining all of the intersections marked 80 dBA, then all of the intersections marked 81 dBA, and so

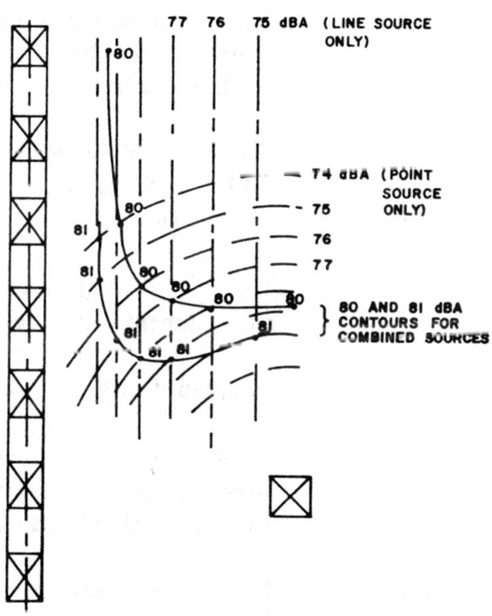

Figure 12-7 Noise contours for combined sources.

on. In the illustration, only the 80- and 81-dBA combined noise contours have been plotted. If noise levels are of interest at greater distances from the noise sources shown, the plotted values may approach background noise levels. In that case, the background noise levels would be combined with the noise levels due to the point and line sources.

GLOSSARY

Absorption	The ability of an acoustical material to absorb sound energy.
Acoustical material	Particular material that is engineered to absorb sound.
Acoustic output	The degree of a noise source in terms of sound power level or sound pressure level. The units are expressed in decibels.
Acoustics	The field of science that includes the study of sound generation and transmission.
Airborne sound	The sound that reaches the point of measurement traveling through the air.
Alias	Sampled data that are equally spaced.
Ambient noise	Noise that is a mixture of different airborne sounds from many sources around the point of measurement.
Amplitude density distribution	The function expressing the fraction of time that pressure, voltage, or another variable dwells in a narrow range.
Analyzer	The combination of a filter system and a device used for evaluating the relative energy that passes through the filter system.

Anechoic	An environment that is very close to a free field.
Anechoic room	A room whose structure absorbs all incident sound. The purpose of such a room is to set up free-field conditions.
Articulation index	An index aiding in properly classifying certain vowel and consonant sounds. It is labeled from 0 to 1 and is a fundamental method of identifying human speech sounds.
Audiogram	A plot of data that indicates the hearing threshold level (HTL) as a function of frequency.
Audiometer	A testing device used for measuring hearing ability and threshold levels.
Aural	That which has to do with the ear or to the sense of hearing.
Autospectrum	Also known as the power spectrum. It is the spectrum whose coefficients of components are presented as the square of the magnitudes.
Background noise	Noise that is picked up in testing that does not originate from the source of interest.
Baffles	Components inserted in the flow path through rectangular silencers, usually to attenuate high-frequency noise.
Center bodies	Elements inserted in the flow path through circular silencers, usually to attenuate high-frequency noise.
Coherence	A measure of a transfer function estimate. It has a value of 1 when the estimate has not been infiltrated by background noise.
Confidence limits	The limits (upper and lower) of a certain range over which a definite percent probability applies.

Critical speed	The speed of a rotating system that applies to a resonance frequency of the system.
Cross correlation	The measurement of similar characteristics of two functions whose displacement is an independent variable, usually time.
Cross spectrum	The measure inside the frequency domain of the similar characteristics of two functions.
Damp	To produce a loss of vibration or oscillatory energy in a mechanical or electrical system.
Damping	The loss of vibrational or oscillatory energy from a mechanical or electrical system.
Data window	The form of a weighing function that is considered as multiplying the data that enter a calculator.
Dead room	A room whose structural materials absorb most or all of any incident sound.
Decibel	A unit of sound pressure-squared level.
Degrees of freedom	A degree of stability related to the number of independent equal terms entering into a distribution.
Diffraction	Any change in the direction of propogation of sound energy waves. Diffraction usually occurs at a boundary discontinuity of an absorptive material.
Diffuse sound field	An area in which the average rate of sound energy flow from a source is equal in all directions away from the source.
Directivity factor	The ratio of the square of the sound pressure to the mean square pressure.

	Used for sound emission in a transducer, this ratio is determined at a specified distance and direction from the source. For an accurate factor the distance from the source must be large so the sound will appear to diverge spherically.
Direct sound field	An area in which a majority of the sound comes directly from the source without any contributions due to reflection.
Displacement	A vector quantity indicating the change of position of a particle, measured from the mean position.
Dissipative silencers	An arbitrary designation for silencers that absorb rather than reflect noise, usually in a broad frequency range.
Dynamic insertion loss	Insertion loss measured under flow conditions in silencers.
Earphone	An electroacoustic transducer that is tightly coupled to the ear.
Effective sound pressure	The root mean square of the instantaneous sound pressures over a given length of time at the measurement point.
Face area	The cross-sectional area of the inlet portion of a silencer.
Face velocity	The velocity of a cross section of gas or liquid at the inlet face of a silencer.
Filter	A device separating the components of an incoming signal by its frequencies.
Flanking transmission	The transmission of sound from a source through a barrier that is not an acoustical material being directly tested. Such a barrier would be a partition.
Folding frequency	The inverse of twice the time interval between sampled values.

Frame	A group of values analyzed as a group.
Frame size	The number of sample points that are in the frame.
Free-sound field	An area in which the effects of the boundaries upon the sound field are irrelevant; a field that is free of boundaries.
Frequency	The rate of time for a full cycle to be completed in a periodic function. The units of frequency are Hertz (Hz).
Gaussian distribution	An amplitude distribution whose histogram is bell shaped. Also known as normal distribution, it pertains in the sound field to stationary acoustic noise that is not periodic.
Hanning	A data window that is used in the form of a time domain of a raised cosine arch. At the beginning and end of the frame, its weight is zero; in the middle, it weighs unity.
Hearing threshold level	The amount (in decibels) of the threshold of audibility of the ear exceeding the standard audiometric threshold.
Histogram	A plot of amplitude density distribution.
Impact	The collision of a moving mass into another mass that is at rest or also in motion.
Insertion loss	The difference in decibels between two sound levels that are measured at the same point in space before and after a muffler or a silencer is inserted between the measurement point and the source of the noise.
Instrument noise	The electrical sound that is generated in the measuring devices of airborne sound.

Isolation	The complement of transmissibility in a steady-state forced vibration.
Jerk	The third derivative of the displacement with respect to time. A vector giving the time rate of change of acceleration.
Just noticeable difference	The amount of difference required so as to be classified that there has been a change in any attribute sound.
Level	The logarithm of the ratio of a quantity to a reference quantity of the same type.
L =	log, (q/q^0)
where:	
L =	level of quantity (units measured in $\log_r r$)
r =	reference ratio and the logarithm base
q =	the quantity under consideration
q_0=	the reference quantity of the same field
Level distribution	An array of quantities that shows noise exposure by giving the duration of time that the sound-pressure level prevailed within each of a set of level intervals.
Line component	A simple tone. It may or may not be a portion of a complex signal.
Live room	A room whose sound absorptive qualities are very poor.
Loudness	A strong auditory sensation. Loudness depends on sound pressure, frequency, and waveform of the stimulating sound.
Loudness contour	A plot of data that shows related values of sound pressure levels and frequency that is needed to produce a specific loudness sensation.

Loudness level — The sound that is equivalent to the median sound-pressure level relative to 0.0002 microbar of several trials evaluated by persons to be of equal loudness at 1,000 Hz. Units are phons.

Loudspeaker — An electrostatic transducer that emanates acoustic power into the surrounding environment. The resultant acoustic waveform is equal to the electrical input.

Mach number — The ratio of the flow speed to the speed of sound in a silencer.

Masking — The magnitude that the threshold of audibility of a noise is increased by the presence of another masking sound within the environment. Units are expressed in decibels.

Mechanical impedance — The impedance obtained from the ratio of force to velocity during simple harmonic motion.

Mel — A unit of pitch. The pitch of a sound that is evaluated to be *n* times that of a one-mel tone is *n* mels.

Microbar — A unit of pressure used in acoustics. It is a dyne per square centimeter.

Microphone — An electrostatic transducer that is stimulated by sound waves, transmitting equivalent electric waves.

Muffler — A different name for a silencer.

Neper — A division on a logarithmic scale denoting the ratio of two similar quantities that are proportional to power or energy. This ratio is converted to nepers by multiplying the logarithm to the base *e* by 0.5.

Noise level — The actual quantity of noise that is measured in specific units such as decibels or volts.

Noise level for airborne sound	The sound level that is the weighted sound pressure level.
Noise reduction	A difference expressed in decibels between space-time average sound pressure levels that are produced in two rooms by any number of sound sources contained in them.
Noise reduction coefficient (NRC)	The average of sound absorption coefficients at frequencies of 250, 500, 1,000 and 2,000 Hertz.
Noys	A dimension used to evaluate the perceived noise level.
Octave	Either the pitch interval between two tones or also the interval between two sounds with a frequency ratio of two.
Octave band	The frequency band whose upper band-edge frequency is double that of the lower band-edge frequency.
One-third octave band	The frequency band that has its upper band-edge frequency at one third the lower band-edge frequency.
Oscillation	The variation with time of the magnitude of a value with respect to a specific reference, when the magnitude is alternated greater and smaller than the reference.
Parallel silencer	A combination of silencers connected in parallel, such as putting one inside the other.
Partition	Any structural component that separates space, such as wall, door, floor, or ceiling.
Peak-to-peak value	The algebraic difference between the two extremes of quantity of an oscillating parameter.

Perceived noise level	The level of noise that is obtained through a calculation process based on an approximation to subjective evaluations of noise magnitude (units are in decibels).
Periodic quantity	An oscillating quantity repeated over certain independent variable increments.
Phon	Units used in denoting loudness levels.
Pink noise	A certain type of sound where the noise-power-per-unit frequency is inversely related to the frequency over a specific range.
Pitch	The characteristic of an auditory sensation in which noises are ordered on a scale that extends from low to high.
Point source	A noise source that sets up a uniform sound field under free field conditions.
Power level	Ten times the logarithm to the base 10 of the ratio of a certain power to a referenced power. Reference power is shown when the level is recorded (units are in decibels).
Presbycusis	Loss of hearing due mainly to aging.
Pressure drop across a silencer	The difference between upstream pressure and downstream pressure at a certain flow rate through the silencer, normally measured in inches of water.
Pure tone	Sound sensation that is characterized by singleness of pitch. Also referred to as a simple tone.
Random noise	An oscillation whose instantaneous magnitude cannot be specified for any point in time.
Rate of decay	The rate at which the sound pressure level decreases after a source has stopped giving off sound energy.

Reactive silencer	An arbitrary designation for silencers that are tuned to reflect, rather than absorb, noise in narrow frequency ranges.
Reflective environment	An environment in which large sound absorptive surfaces are present.
Resonance frequency	The frequency at which resonance takes place.
Reverberation	Unyielding sound in an enclosed area as a result of repeated reflection after the source has stopped giving off sound energy.
Reverberation room	A room designed to approximate the reverberant sound field as a diffuse sound field.
Reverberant sound field	An enclosed area where most or all of the present sound is constantly reflected off its perimeters.
Salum	A unit of sound absorption measure.
Self-noise	The noise generated by air flowing through a silencer.
Series silence	A combination of silencers connected in series, such as one located downstream from the other.
Side branches	Sound-absorbing elements connected on the sides of silencers, normally for the purpose of attenuating low-frequency noise.
Silencer	A device that allows a functional work flow while preventing the transmission of noise.
Sociocusis	A specific condition that increases the threshold hearing level due to exposure to noise that is not related to the social environment exclusive of occupational

noise exposure, physiologic changes over time, and otologic diseases.

Sone — The tiniest increment of a psychophysical scale indicating the intensive attribute of complex sounds. A unit of loudness.

Sonics — The science of sound in analysis and processing.

Sound — Auditory sensation developed by the oscillations in pressure, stress, particle velocity, and particle displacement in a medium containing internal forces, or the superposition of such propogated alterations.

Sound absorption — A process by which one eliminates sound energy.

Sound energy — Energy that enters a medium where sound travels. It consists of potential energy in the form of deviations from static pressure and kinetic energy in the form of particle velocity.

Sound-energy density — The amount of sound energy per unit of volume.

Sound intensity — The average rate of sound-energy flow in a specific direction divided by the area, perpendicular to that specific direction, through or toward which it travels.

Sound power — Symbolized by *W* and having the units of watts, it is the rate at which acoustic energy is radiated.

Sound pressure — A fluctuating pressure that is superimposed on the static atmospheric pressure. It is denoted by *P* and has the units of dyne/cm^2 or Newton/m^2.

Sound transmission class (STC) — A rating system using single figures that is engineered to give a preliminary

	properties of a partition or preliminary rank of a series of partitions.
Sound trap	Another name for a silencer.
Tone	An auditory sensation having pitch.
Transducer	A device that is actuated by waves from transmission systems, and in turn transfers waves to other transmission systems.
Vibration	Oscillation where the quantity is a parameter that defines the motion of a mechanical system.
Vibration isolator	A support that isolates a system from any steady-state excitation.
Vibration meter	An instrument that can measure the displacement, velocity, or the acceleration of a vibrating body.
Volume flow velocity	The rate at which a specific volume flows through a silencer, usually cubic feet per minute.
Weighting	A specific frequency response given in a sound level meter.

INDEX

A

A-weighted sound level 135
abatement of hazards 26
absorption 174
acceleration 99
accelerometer 107
accuracy 157
acoustic output 174
acoustic reflex 4
acoustical barriers 2
acoustical ceiling 91
acoustical foams 80
acoustical material 174
acoustical measurement 32, 151
acoustical panel 82
acoustical seal 59
acoustics 32, 174
adjustable dampers 114
administration 13, 24
administrative controls 1, 20
age groups 4
air springs 113
air-conditioning 131
air-suspension systems 114
airborne sound 174, 181
allowable sound pressure 159
alternate programs 160
ambient noise 174
amplifier 136
amplitude analysis 155
amplitude density distribution 174
amplitude distribution analyzers 153
amplitude relationships 100
amplitude response 41
analyzer 174
anechoic 175
anechoic room 175
ANSI standard 156
antivibration block 128
antivibration mounting 119
applications for isolators 128
articulation index 175
assembly errors 49
assembly of equipment 50
attenuation 59, 63
attenuation graph 63
attenuation levels 61
attenuation quality 59
attenuator 136
audiogram 8, 175
audiometer 175
audiometric test booths 158
audiometric testing 156
aural 175
automatic audiometers 156
autospectrum 175

B

background noise 175
baffles 175

ballistocardiogram 8
barrier installation 73
barriers 68, 165
belt drives 47
blocks 128
blood vessels 7, 8
bone conduction 58
break periods 12
bumper isolator 124
bumper-type vibration 123

C

calibration 150
ceiling construction 55
ceilings 79
center bodies 175
characteristics of speech 7
circular duct silencer 90
cleaning of equipment 50
cochlea 3
codes 163
coherence 175
combined noise 173
combined noise sources 170
communication 64
community noise 163
community reaction 40
complaints of violations 24
complex noise 42
compliance 163
compliance officers 25
compressors 46, 132, 140
concrete foundations 116
condensers 132, 140
confidence limits 175
contesting a citation 25
continuous spectrum 43
contour mapping 172
control measures 19
control methods 49
controlling vibration 111
cooling towers 132, 140
cork 116
cotton 61
Coulomb damping 115
critical speed 176
cross correlation 176
cross spectrum 175
curtain 53

D

damp 176
dampening vibration 20
damping 49, 50, 176
damping compound 89
damping materials 50
dangerous properties 4
data window 176
dead room 176
deafness 4
decibel 38, 176
decrease turbulence 50
degrees of freedom 176
design standards 18
detection of sound 32
diffraction 176
diffuse sound field 176, 177
directivity factor 176
displacement 99, 177
displacement measurement 104
dissipative silencers 177
dosimeters 151, 156, 161
drive belts 49
drywall partition 55
duct silencer 90
ductwork 73
duties of employers 23
dynamic insertion loss 177
dynamic transmission loss 135

E

ear 2
ear canals 62
ear plugs 49, 66
ear protection 59

ear protection program 66
ear protective devices 57
ear protectors 60, 64, 66
eardrum 3
earphone 177
effective sound pressure 177
effects of noise 4
effects of reflection 36
effects on the human body 7
elastic connections 119
electric motor 154
electrode microphone 143
element transducer 106
employer-employee relations 12
enclosures 49, 53, 68
engineering control measures 19
engineering controls 1, 44
equipment noise sources 45
evaporators 133, 140
excessive vibration 48
exhausts 47
exposure duration 6

F

fabric backing 53
face area 177
face velocity 177
fans 140
far field 165
faulty installation 46
federal supervision 11
fiber-loaded vinyl 88
filter 152, 177
flanking transmission 177
flexible polyurethane 84
foam 63
folding frequency 177
frame 178
frame size 178
free-sound field 178
frequency 178
frequency analyzers 152
frequency and wavelength 35
frequency response curves 108
frequency spectra 59

G

gaussian distribution 178
gear problems 49
gears 47
glass barrier 53
graphic recording 153

H

hanning 178
harmonic oscillation 34
harmonic vibration 32, 93
hazardous jobs 12
health standards 24
hearing aid 5
hearing conversation 67
hearing disorder 4
hearing threshold level 178
high-frequency vibration 99
histogram 178
horizontal standards 13
horizontal vibrations 122
human ear 2

I

imminent dangers 27
impact 178
impact noise 19
impaired hearing 5
implementation 28
in-plant testing 160
industrial environments 59
industrial mountings 119
industrial sound attenuation 69
industrial sources 163
inertia blocks 112
inner ear 3, 58
inner-ear protection 61

insertion loss 135, 178
inspections 25
installation principles 117
instrument noise 178
instrumentation 141
intelligibility scores 65
interim standards 24
interpolation 162
inverse distance law 168
inverse square law 36, 165
isolating noise 52, 53
isolating noise sources 20
isolating operator 20
isolation 50, 179

J

jerk 179
job schedules 2
joist systems 78

L

lagging 49
lead 68
lead materials 69
lead plenum 71
lead powder 53
lead-loaded fabric 56
lead-loaded vinyl 89
lead-plywood 70
level 179
leveling screw 123
levels of displacement 102
line component 179
line sources 168
line spectra 42
live room 179
loudness 39, 179
loudness contour 41, 179
loudness level 133, 180
loudspeaker 180

M

Mach number 180
machines substitution 46
maintenance 44
maintenance of equipment 20
malleus 3
mapping 162
masking 180
measurement techniques 136
measuring displacement 98
mechanical impedance 180
medical studies 1
mel 180
metal ducts 137
microbar 180
microphone 136, 141
middle ear 2, 3
minimum audible field 4
misalignment 49
mobile van 160
motor-driven equipment 111
mounting equipment 2
mounting of sleeves 125
mounts 111
muff-type protectors 62
muffler 180
muffs 49
multiple-chamber plenum 139

N

natural rubber 112
nature of sound 32
near field 165
neper 180
noise barrier 68
noise contours 172
noise criteria 133
noise criteria curves 134
noise enforcement 30
noise exposure levels 136
noise interpolator 167
noise level 180

noise level specifications 20
noise levels 1
noise measurements 141
noise pollution level 135
noise power 35
noise reduction 137, 181
noise reduction coefficient 181
noise regulation 11
noise spectrum 5
noise-exposed group 5
noise-level mapping 164
noise-level ranges 50
noise-level specifications 52
noys 181
nylon 87
nylon gears 87

O

objective measurements 39
Occupational Safety and Health Act 11
octave 181
octave analyzers 153
octave band 6, 181
octave filter set 153
office areas 54
office construction 70
one-third octave band 181
operator isolation 54
ordering new equipment 20
oscillation 181
OSHA Standards 13

P

package attenuators 137
pads 111
parallel silencer 181
partition 181
peak-to-peak value 181
penalties for violations 26
penalty 25
perceived noise 182
performance standards 18
periodic function 43, 93
periodic quantity 182
permanent standards 24
permanent threshold shift 6
permissible noise exposures 6, 18
personal ear protectors 62
personal safety 57
phon 182
physical examinations 12
physical problems 9
piezoelectric accelerometer 106, 107
piezoelectric microphones 143
pink noise 182
pipes and wires 72
piping 140
pitch 182
plenum barriers 70
plenums 49
plotting line source 169
plugs and muffs 61
plywood lead 70
pneumatic equipment 46
point source 166, 182
policy 28
polyester film 82
polymer film 82
polymer-coated foam 83
polyurethane foam 63, 69, 84
power level 182
presbycusis 182
pressure drop 182
procedures 28
process substitution 47
production equipment 55
propagation 32
protection against harassment 27
protective device 58
protective equipment 1
pumps 140
pure tone 39, 182
PVC film 88

R

random noise 182
random vibration 95
rate of decay 182
reactive silencer 183
realignment 50
recordkeeping requirements 27, 29
records 158
recruitment of loudness 4
rectangular air ducts 73
reducing sound transmission 20
reduction of flow 50
reflective environment 183
relation to distance 7
resonance frequency 183
responsibilities of employees 12
responsibilities of employers 12
reticulated polyurethane 87
reticulation 81
return grille 92
reverberant field 165
reverberant sound field 183
reverberation 183
reverberation room 183
rights of employees 23
rotating equipment 50
rotating personnel 2
rubber 116
rubber isolators 48, 112

S

salum 183
sanitation 64
Secretary of Labor 13
self-noise 183
sensitivity for hearing 6
sensitivity to vibration 9
series silence 183
shaft coupling 126
sheet lead 69
shroud 53
silencer 90, 183
silencers or mutes 49
simple harmonic motion 99
sinusoidal signal 93
sinusoidal vibration 33, 42
sliding parts 115
sociocusis 183
sone 184
sonics 184
sound 184
sound absorption 184
sound diffuseness 37
sound energy 184
sound field 35
sound intensity 184
sound isolation 55
sound level meter 136, 146, 152
sound power 184
sound pressure 184
sound pressure level 37
sound propagation 34
sound spectrum 43
sound transmission class 184
sound trap 185
sound waves 32
sound-absorbing foams 86
sound-damping 53
sound-energy density 184
sound-wave propagation 32
sources of vibration 96
speech intelligibility 65
speech interference level 133
spherical radiation 36
spherical wave fronts 35
spring systems 113
standards 163
statistics 27
steel springs 48, 117
steel springs 48
stiffness 52
strain gauge 105
structure-borne vibrations 55
substitution of machines 19
substitution of processes 20
supervision 67

suppressor 90
surveys 1
suspended ceiling 70, 92
systems design 54

T

temperature resistance 52
temporal distribution 6
temporary threshold shift 6
tolerance levels 60
tolerances 4
tone 185
transducer 185
transfer grilles 91
transient vibration 96
transillumination 7
transmission loss 135
two noise sources 171
types of vibration mounts 116

U

union examiner 12
urethane foams 80

V

variable noise 19
variances 26
vasoconstriction 8
velocity 99
velocity measurement 106
velocity of propagation 33
ventilating systems 131
vertical standards 13
vibration 185
vibration 8, 93
vibration control 111
vibration damping 47
vibration effects 97
vibration isolator 116, 185
vibration isolator mats 48
vibration measurement 97
vibration meter 109, 185
vibration on man 8
vibration plates 127
vibration tranducers 104
vibration-related problems 9
vibratory motion 9
vibrograph 104
vinyl 53
viscoelastic materials 49
volume flow velocity 185

W

Walsh-Healey Public Controls Act 21
weighting 185
weighting networks 42
welding 47
work breaks 2
working areas 1
worn parts 50